Editors

Lecture Notes in Chemistry

Edited by G. Berthier M. J. S. Dewar H. Fischer
K. Fukui G. G. Hall H. Hartmann H. H. Jaffé J. Jortner
W. Kutzelnigg K. Ruedenberg E. Scrocco

32

H. F. Franzen

Second-Order Phase Transitions and the Irreducible Representation of Space Groups

Springer-Verlag
Berlin Heidelberg New York 1982

Author

H. F. Franzen
Dept. of Chemistry, Iowa State University
Ames, IA 50011, USA

ISBN 978-3-540-11958-6
DOI 10.1007/978-3-642-48947-1

ISBN 978-3-642-48947-1 (eBook)

2152/3140-543210

PREFACE

The lecture notes presented in this volume were developed over a period of time that originated with the investigation of a research problem, the distortion from NiAs-type to MnP-type, the group-theoretical implications of which were investigated in collaboration with Professors F. Jellinek and C. Haas of the Laboratory for Inorganic Chemistry at the University of Groningen during the 1973-1974 year. This distortion provides the major example that is worked through in the notes.

The subject matter of the notes has been incorporated in part in the lectures of a course in Solid State Chemistry taught several times at Iowa State University, and formed the basis of a series of lectures presented at the Max-Planck Institute for Solid State Research in Stuttgart during 1981-1982, and as part of a Solid State Chemistry course taught during the spring of 1982 at Arizona State University in Tempe. I wish here to express my gratitude to the Max-Planck Institute for Solid State Research and to Arizona State University for the opportunity and support they provided during the time I was developing and writing the lecture notes of this volume.

I wish also to thank the many colleagues and students who have offered comments and suggestions that have improved the accuracy and readability of the notes, and who have provided stimulation through discussion of the ideas presented here. I am especially indebted to Professors C. Haas and F. Jellinek, who helped me through the difficult initial learning process, to Dr. J. Folmer with whom I spent many rewarding hours discussing applications and implications of Landau theory, and to Professors H.-G. von Schnering and A. Simon of the MPI, Professor H. Bohm of the University of Munster and Professor M. O'Keeffe of Arizona State University for their enthusiastic and critical participation in the lectures for which these notes formed the basis.

Finally, I want to thank Shirley Standley for her substantial efforts in the preparation of the manuscript.

TABLE OF CONTENTS

SPACE LATTICE SYMMETRY

I. INTRODUCTION

Some properties of crystalline solids, such as the directions and symmetry of diffraction (X-ray, neutron, electron), the anisotropy of transport phenomena (electrical, thermal conduction, matter diffusion), and the anisotropy of thermal expansion and interactions of crystals with optical radiation are directly related to the underlying three-dimensional periodicities that characterize many crystalline materials. It is therefore important to understand such periodicity in order to better understand the phenomenology of the interactions of crystalline solids. It is also necessary to understand three-dimensional periodicity as a basis for development of the abstract theory of symmetry in crystalline solids (the theory of space groups and their representations). The three-dimensional periodicities are conveniently discussed in terms of space lattices, which are three-dimensional spatial arrays of discrete points which correspond to the translational symmetries of solids. The theory of space lattices, developed below, is the theory of the allowed symmetries of the periodicity of three-dimensionally crystalline solids.

II. TRANSLATIONAL PERIODICITY

Crystalline solids with <u>three-dimensional periodicity</u> are characterized by sets (signified by { }) of <u>translational symmetry operations</u>, $\{T_{mnp}\}$ that are linear combinations of noncoplanar basis vectors, $\vec{a}$, $\vec{b}$, $\vec{c}$:

$$\vec{T}_{mnp} = m\vec{a} + n\vec{b} + p\vec{c},$$

where m, n and p are integers and the lengths of and angles between the basis vectors vary from one solid to another and, subject to restraints to be discussed below, also with the thermodynamic state of a given crystalline solid. The vectors, $\vec{T}_{mnp}$, are between all points that have symmetrically equivalent environments.

A point in space within a crystal can be defined relative to an arbitrary origin by $\vec{r} + \vec{T}_{mnp}$ where

$$\vec{r} = x\vec{a} + y\vec{b} + z\vec{c}$$

with x, y and z between zero and one. All such points with different integral values of m, n and p and fixed x, y and z are points which are indistinguishable by virtue of their environments. The point defined by $\vec{r}$ is frequently designated by x, y, z and the vector $\vec{T}_{mnp}$ by m, n, p and the above can be restated, "the points x, y, z and x+m, y+n, z+p are equivalent by symmetry in a crystalline solid exhibiting three-dimensional periodicity".

The set of all points generated by the termination of the vectors $\vec{r} + \vec{T}_{mnp}$ when m, n and p take on all integral values for a given x, y, z is called a <u>space lattice</u>. The space lattice has the property the each lattice point is in a symmetrically equivalent environment. The origin of the space lattice is the point x, y, z, and this point can be chosen with any triple of values between zero and one, i.e., at any point within a <u>unit cell</u>.

The three vectors $\vec{a}$, $\vec{b}$ and $\vec{c}$ thus define a unit cell of the lattice; the volume of this cell is given by

$$V_{cell} = \vec{a} \cdot \vec{b} \times \vec{c}.$$

This is the volume per lattice point of the lattice or is the <u>primitive</u> cell volume. It is sometimes advantageous to designate cells which describe the symmetry of a lattice (to be discussed below), but for which there exist translational symmetry operations that are not included in the set {Tmnp}. Such cells have more than one lattice point per cell and it is understood

that other translational symmetry operations are included in the complete
set. For example if it is said that a lattice is <u>body-centered</u> then it is
understood that the translational symmetry operations

$$\{T_{mnp} + \frac{\vec{a} + \vec{b} + \vec{c}}{2}\}$$

must be added to $\{T_{mnp}\}$ to obtain the complete set, i.e., each unit cell has,
as well as a 1/8 share of each lattice point at each of 8 corners, a lattice
point in the center of the body of the cell. Such body-centered cells
contain two lattice points per cell volume. It is then also possible to
define a primitive cell, for example by letting

$$\vec{a}_p = \frac{1}{2}(\vec{a} + \vec{b} + \vec{c}),$$

$$\vec{b}_p = \frac{1}{2}(\vec{a} + \vec{b} - \vec{c}),$$

$$\vec{c}_p = \frac{1}{2}(\vec{a} - \vec{b} + \vec{c}),$$

and this cell, labeled p, would contain one lattice point per cell. However,
the symmetry of the lattice would not generally be fully described by the
conventional description of the symmetry of the primitive cell and in this
case the centered cell would be chosen. Other centerings, besides body-
centering, are end-centering (e.g. $(\vec{a} + \vec{b})/2$ for C centering, $((\vec{b} + \vec{c})/2$
for A centering, etc.) and face centering $((\vec{a} + \vec{b})/2$ and $(\vec{b} + \vec{c})/2$ and
$(\vec{a} + \vec{c})/2)$.

One type of symmetry of space lattices is the proper rotational
symmetry of the lattice. A lattice has proper rotational symmetry when there
exists a line such that rotation by some minimum angle, α, about the line
causes the appearance of the lattice to be indistinguishable from what it was
prior to the rotation. All rotational symmetry operations about the axis are

then rotations by angles $m\alpha$, where m is an integer, and since rotation by 360° about the axis is a symmetry operation it follows that there exists an integer n such that $n\alpha = 360°$. The rotational symmetry <u>element</u> (the line about which the lattice is rotated) is then called an n-fold proper axes and the rotational symmetry <u>operations</u> corresponding to this axis are C_n, C_n^2, C_n^3, ..., $C_n^n = \varepsilon$, where rotation by 360° is symbolized ε because it is equivalent to the identity operation.

For example if $\alpha = 60°$ then the axis is a 6-fold proper axis and the corresponding operations are C_6 (rotation by 60°), $C_6^2 = C_3$ (rotation by 120°), $C_6^3 = C_2$ (rotation by 180°), $C_6^4 = C_3^2$ (rotation by 240°), C_6^5 (rotation by 300°) and $C_6^6 = \varepsilon$. Note that the operations of a 6-fold axis include those of a 3-fold and 2-fold axis, i.e., a 6-fold axis is necessarily also a 3-fold and a 2-fold axis.

In order to discuss the proper rotational symmetries of space lattices it is convenient to initially treat the proper rotational symmetries of plane lattices. Since any two vectors are necessarily coplanar any lattice defined by the vectors $\{\vec{T}_{mnp}\}$ can be generated by the integral repitition of the plane lattice defined by $\{\vec{T}_{mn}\}$, where $\vec{T}_{mn} = m\vec{a} + n\vec{b}$, according to the stacking vector $\vec{c}$. It follows that the proper rotational symmetries allowed for three-dimensional lattices are those allowed for two-dimensional lattices. In order to consider these rotational symmetries it is necessary to make use of the following result: If a plane lattice has an n-fold symmetry axis then it has an n-fold axis through every lattice point. The proof of this statement is illustrated in Fig. 1. The open circle represents the location of the symmetry axis-lattice plane intersection. The two black dots represent a lattice point and one

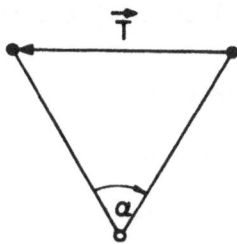

Fig. 1. Open circle: point of intersection of rotational axis $\left(\frac{360}{\alpha}\text{-fold}\right)$ with plane of figure. Black dots: two lattice points related by rotation by α. $\vec{T}$ is the implied translational symmetry operation

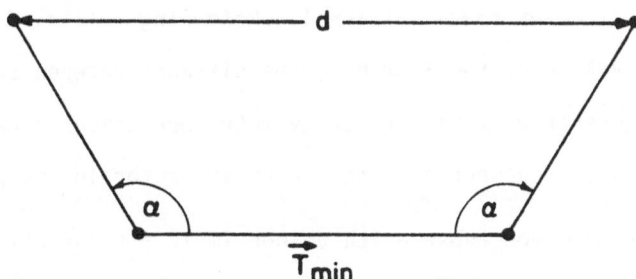

Fig. 2. Two lattice points related by the shortest vector in the plane $(\vec{T}_{min})$ and two additional lattice points (separated by distance d in the direction $\vec{T}_{min}$) implied by rotation by α and $-\alpha$.

related to it by the rotational symmetry operation C_n ($n = \frac{360}{\alpha}$). It follows

that $\vec{t}$ is a translational symmetry operation. If, as supposed, C_n is a

symmetry operation of the lattice and $\vec{t}$ is therefore also a symmetry

operation, then the combined operation $C_n \mid \vec{t}$, rotation by α about the open

circle followed by translation by $\vec{t}$, is also a symmetry operation. However

$C_n \mid \vec{t}$ is equivalent to rotation by α about a lattice point, completing the

proof.

With this result it can now be shown that translational periodicity

significantly restricts the allowed proper rotational symmetries of plane

lattices (and therefore of space lattices). Referring to Fig. 2, we inquire

about the allowed values of the length d, the distance between two lattice

points which are generated by the proper symmetry operations $\pm C_n$ from two

lattice points that are connected by the shortest vector in the plane ($\vec{t}_{min}$).

The allowed values of d are those which do not imply a translational symmetry

operation of length less than $|\vec{t}_{min}|$, therefore the allowed values of d are 0,

$|\vec{t}_{min}|$, $2|\vec{t}_{min}|$ and $3|\vec{t}_{min}|$. From these d values the allowed values of α

follow, namely 60°, 300°, 90°, 270°, 120°, 240°, 180° and, trivially, 360°.

Thus the only allowed axes are 6-fold, 4-fold, 3-fold, 2-fold and, of course,

1-fold. All plane lattices exhibit 2-fold symmetry (both $\vec{t}$ and $-\vec{t}$ are

translational symmetry operations), it therefore follows that if a plane

lattice has 3-fold symmetry it has both 3-fold and 2-fold symmetry through

the lattice points, and therefore has 6-fold rotational symmetry. Thus the

allowed proper rotational axes through the lattice points of the plane

lattices are 2-fold, 4-fold and 6-fold axes.

Another symmetry of plane lattices that can be readily recognized is reflection through a plane perpendicular to the plane lattice. Recognizing the proper rotational symmetries and the reflection symmetries allows the characterization of the allowed plane lattices shown in Fig. 3. Except for the centered rectangular lattice, the lattices shown follow directly from the proceding discussion. The centered rectangular plane lattice has a nonprimitive cell (two lattice points per cell) which is used for descriptive purposes to call attention to the symmetry that results when the primitive parallelogram cell fulfills the special condition $|\vec{a}| = |\vec{b}|$.

The notation used to describe the lattice symmetries specifies two m's to indicate that there are two different mirror planes which, although they are not equivalent by symmetry (i.e., no symmetry operation carries one into the other) nonetheless are required to occur together by the combined operation of one vertical mirror and the axis. Fig. 4 illustrates the case for a vertical mirror and a 2-fold axis (rectangular lattice symmetry).

The lattices shown in Fig. 3 exhaust the possibilities for plane lattices. This follows from the limitation on the allowed rotational symmetries and from the fact that setting $|\vec{a}| = |\vec{b}|$ in the rectangular case leads to the square lattice, which is already included. It is natural to inquire why there is no centered square or centered hexagonal lattice. The answers are that a centered square lattice is equivalent to a smaller primitive square lattice, whereas an attempt to construct a centered hexagonal lattice destroys the 6-fold symmetry of the lattice.

III. SYMMETRIES OF PLANE LATTICES

As discussed above, all plane lattices have two-fold axes through lattice points and, by their nature, all plane lattices exhibit translational symmetry operations, $\vec{t}_{mn}$. The combined operation of a C_2 followed by a

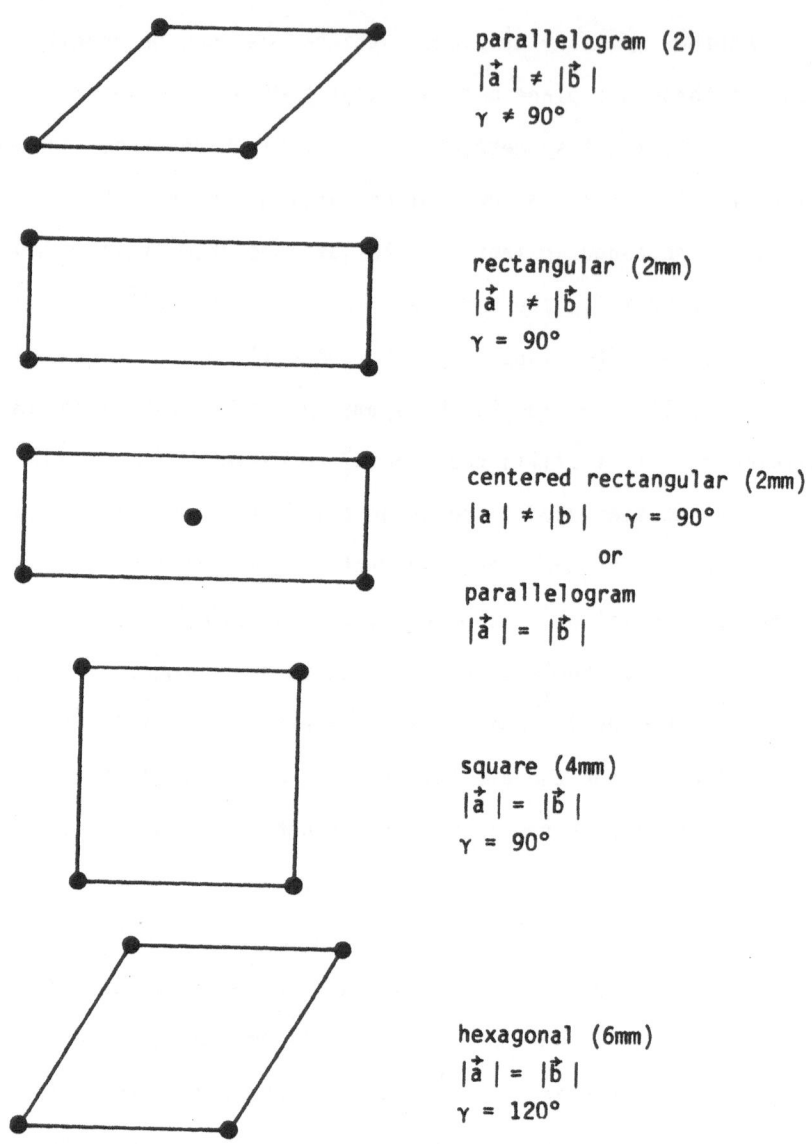

parallelogram (2)
$|\vec{a}| \neq |\vec{b}|$
$\gamma \neq 90°$

rectangular (2mm)
$|\vec{a}| \neq |\vec{b}|$
$\gamma = 90°$

centered rectangular (2mm)
$|a| \neq |b|$ $\gamma = 90°$
          or
parallelogram
$|\vec{a}| = |\vec{b}|$

square (4mm)
$|\vec{a}| = |\vec{b}|$
$\gamma = 90°$

hexagonal (6mm)
$|\vec{a}| = |\vec{b}|$
$\gamma = 120°$

Fig. 3. The five plane lattice types.

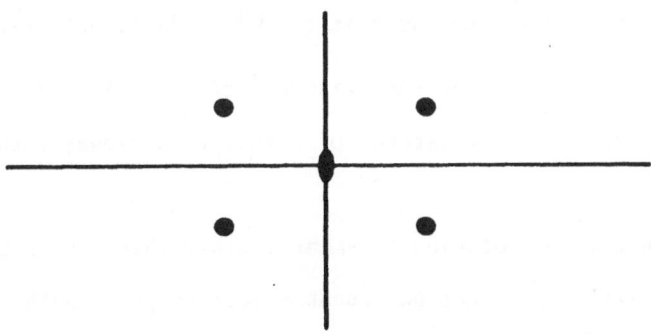

Fig. 4. The four lattice points follow from the two-fold rotation and the reflection operation of one of the mirror planes, the operations of the second mirror plane follow from the combined operations.

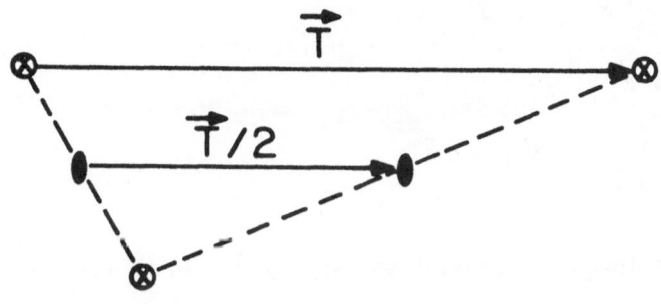

Fig. 5. Demonstration that $C_2 \mid \vec{T}$ is equivalent to a C_2 about $\vec{T}/2$.

translation perpendicular to the axis $(C_2 \mid \vec{t}_{mn})$ is illustrated in Fig. 5. This figure demonstrates that the existence of a 2-fold axis through a lattice point implies the existence of 2-fold axes midway between all pairs of lattice points.

In the case of rotational symmetry other than 2-fold the combination of rotational symmetry operations about a lattice point with translational symmetry operations also results in <u>implied</u> rotational symmetries through points which are not lattice points. However the location and nature of the implied axes must be worked out for each case individually. For the purpose of developing a general scheme for determining the nature and location of implied rotational axes some concepts of operator algebra are useful.

The rotation of a point (x,y) in the a-b plane can be represented by a 2x2 matrix which is symbolized by β:

$$\beta \begin{pmatrix} x \\ y \end{pmatrix} = \begin{pmatrix} x' \\ y' \end{pmatrix} .$$

For example, if $\beta = C_{2z}$

$$\begin{pmatrix} \bar{1} & 0 \\ 0 & \bar{1} \end{pmatrix} \begin{pmatrix} x \\ y \end{pmatrix} = \begin{pmatrix} \bar{x} \\ \bar{y} \end{pmatrix} .$$

The rotation followed by translation, say by $\vec{a}$, can be represented by

$$\begin{pmatrix} \bar{1} & 0 \\ 0 & \bar{1} \end{pmatrix} \begin{pmatrix} x \\ y \end{pmatrix} + \begin{pmatrix} 1 \\ 0 \end{pmatrix} = \begin{pmatrix} \bar{x} \pm 1 \\ y \end{pmatrix}$$

and, as shown above (Fig. 5) this operation corresponds to a C_{2z} operation about $\vec{a}/2$. It is possible to combine the translation and rotation in a single 3x3 matrix:

$$
\begin{pmatrix} \bar{1} & 0 & 1 \\ 0 & \bar{1} & 0 \\ 0 & 0 & 1 \end{pmatrix} \begin{pmatrix} x \\ y \\ 1 \end{pmatrix} = \begin{pmatrix} \bar{x} + 1 \\ \bar{y} \\ 1 \end{pmatrix}
$$

where the last rows of the 3x3 and 3x1 matrices are formally 001 and 1, respectively, so that matrix multiplication yields the correct result. It is worthwhile to note that rotation by C_{2z} about $\vec{a}/2$ is equivalent to the sequence of operations: translation by $-\vec{a}/2$, rotation by C_{2z}, translation by $\vec{a}/2$, i.e., can be represented by

$$
\begin{pmatrix} 1 & 0 & 1/2 \\ 0 & 1 & 0 \\ 0 & 0 & 1 \end{pmatrix} \begin{pmatrix} \bar{1} & 0 & 0 \\ 0 & \bar{1} & 0 \\ 0 & 0 & 1 \end{pmatrix} \begin{pmatrix} 1 & 0 & -1/2 \\ 0 & 1 & 0 \\ 0 & 0 & 1 \end{pmatrix},
$$

which multiply to yield $\begin{pmatrix} \bar{1} & 0 & 1 \\ 0 & \bar{1} & 0 \\ 0 & 0 & 1 \end{pmatrix}$, as they should.

It was our purpose here to discover a means for determining what any proper rotational symmetry operation through a lattice point of a plane lattice implies regarding other symmetry axes. As an example we consider the C_4 through the lattice points of the square lattice. Since rotation by 90° about the origin is represented by

$$
\begin{pmatrix} 0 & \bar{1} \\ 1 & 0 \end{pmatrix} ,
$$

we can write for a C_4 at $x = t_1$, $y = t_2$,

$$
\begin{pmatrix} 1 & 0 & t_1 \\ 0 & 1 & t_2 \\ 0 & 0 & 1 \end{pmatrix} \begin{pmatrix} 0 & \bar{1} & 0 \\ 1 & 0 & 0 \\ 0 & 0 & 1 \end{pmatrix} \begin{pmatrix} 1 & 0 & -t_1 \\ 0 & 1 & -t_2 \\ 0 & 0 & 1 \end{pmatrix} = \begin{pmatrix} 0 & \bar{1} & t_1+t_2 \\ 1 & 0 & t_2-t_1 \\ 0 & 0 & 1 \end{pmatrix}
$$

and equate this with the matrix representing a 4-fold rotation operation about the origin plus translation by $m\vec{a} + n\vec{b}$:

$$
\begin{pmatrix} 0 & \bar{1} & t_1+t_2 \\ 1 & 0 & t_2-t_1 \\ 0 & 0 & 1 \end{pmatrix} = \begin{pmatrix} 0 & \bar{1} & m \\ 1 & 0 & n \\ 0 & 0 & 1 \end{pmatrix}
$$

i.e., $t_1+t_2 = m$ and $t_2-t_1 = n$, relations which fix the location (t_1, t_2) of the C_4 implied by C_4 through the origin and $\vec{T}_{mn}$. For example if translation by $\vec{a}$ is chosen then $m = 1$ and $n = 0$ and $t_2 = t_1 = 1/2$. It follows that a square lattice necessarily has 4-fold axes through the centers of the unit cells, a fact that is obvious upon inspection of such a lattice. The same method applied to a hexagonal lattice (Appendix 1) leads to the conclusion that there are 3-fold axes through the points $1/3$, $2/3$ and $2/3$, $1/3$ of the plane hexagonal cells. The resulting symmetry axes are diagrammed for all plane lattices in Fig. 6.

IV. SPACE LATTICES FROM STACKING OF PLANE LATTICES

       With the knowledge of the plane lattice types and their proper rotational symmetries the possible space lattices can be obtained by considering stacking of the plane lattices with various choices of stacking vector, $\vec{c}$. As with the plane lattices, the space lattices are characterized according to their symmetries. Therefore the approach to the consideration of allowed space lattice types is to examine the ways in which the plane lattices can be stacked so as to preserve axial symmetries. For example, if plane lattices with hexagonal symmetry are stacked such that

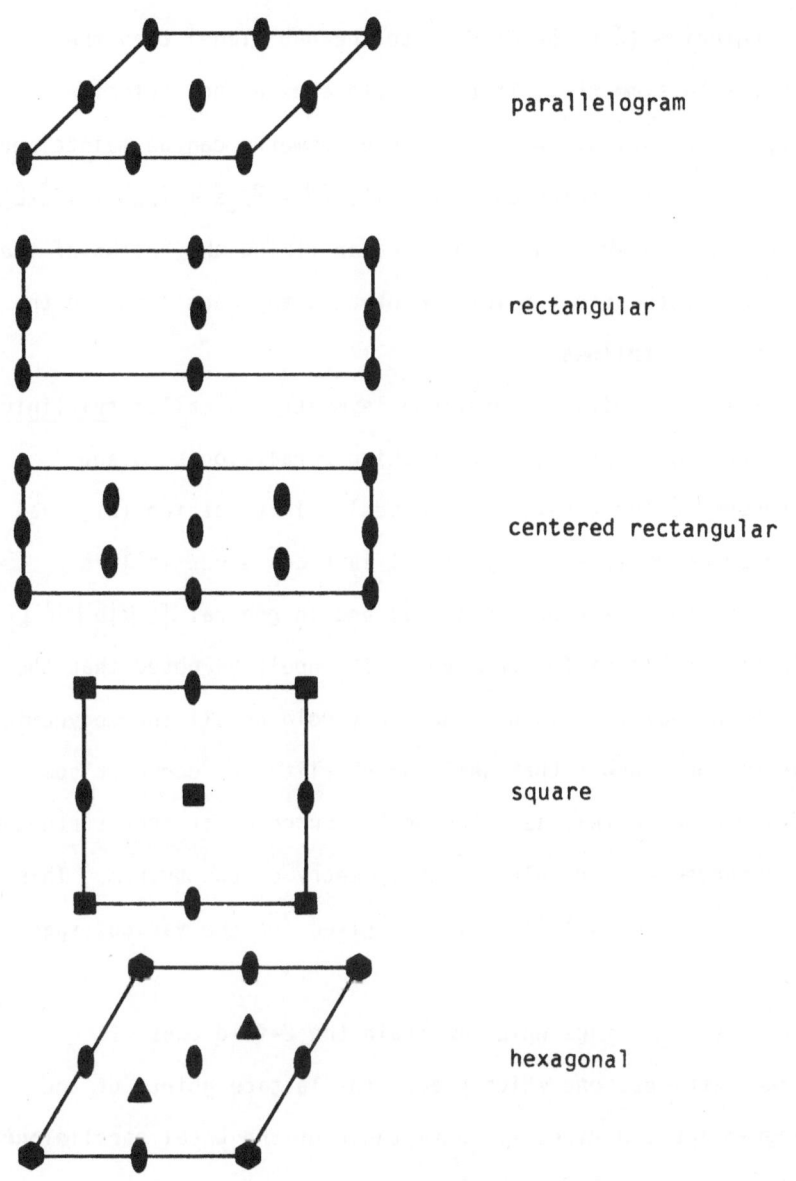

parallelogram

rectangular

centered rectangular

square

hexagonal

Fig. 6. The rotational symmetry axes of the plane lattice types.

their 6-fold axes coincide ($\vec{c}$ perpendicular to the a-b plane) then the space lattice has 6-fold symmetry. If the 6-fold axes do not coincide then the 6-fold symmetry is lost, however 3-fold symmetry can be maintained if the 6-fold and 3-fold axes coincide, i.e., if: ($\vec{c} = \frac{2}{3}\vec{a} + \frac{1}{3}\vec{b} + \tau(\vec{a}x\vec{b})$ or $\vec{c} = \frac{1}{3}\vec{a} + \frac{2}{3}\vec{b} + \tau(\vec{a}x\vec{b})$ where τ is some scalar fixing the height of the stacking layer). The systematic application of this approach leads to the 14 Bravais lattice types, as follows.

A lattice which exhibits no rotational symmetry is called triclinic, and this lattice type can be produced by stacking parallelogram plane lattices with a general $\vec{c}$ (no two-folds coincide). This lattice (Fig. 7a) has only inversion symmetry (a symmetry of all lattices since if $\vec{t}$ is a translational symmetry operation so too is $-\vec{t}$) and in general $|\vec{a}| \neq |\vec{b}| \neq |\vec{c}|$ and $\alpha \neq \beta \neq \gamma$ and all angles differ from 90°. It should be noted that the significance of the inequalities is not that they hold at all thermodynamic states of the system, but rather that while an equality may occur at some isolated thermodynamic state this is a "chance" occurrence at that state and is not generally observed as a result of the symmetry of the system. This remark is generally valid for all inequalities given for the various space lattice types.

There are three stackings which maintain the 2-fold axes of parallelogram plane lattices, one which places the lattice points of the stacked parallelogram lattice directly above those of the basal parallelogram lattice (Fig. 7b), one which places the lattice points of the stacked lattice above the midpoints of the edges (Fig. 7c) and one which places the lattice points of the stacked layer above the center of the cell (Fig. 8). All three of the resulting lattices exhibit 2/m (a two-fold axis perpendicular to a

Fig. 7. The 14 Bravais Lattice Types.

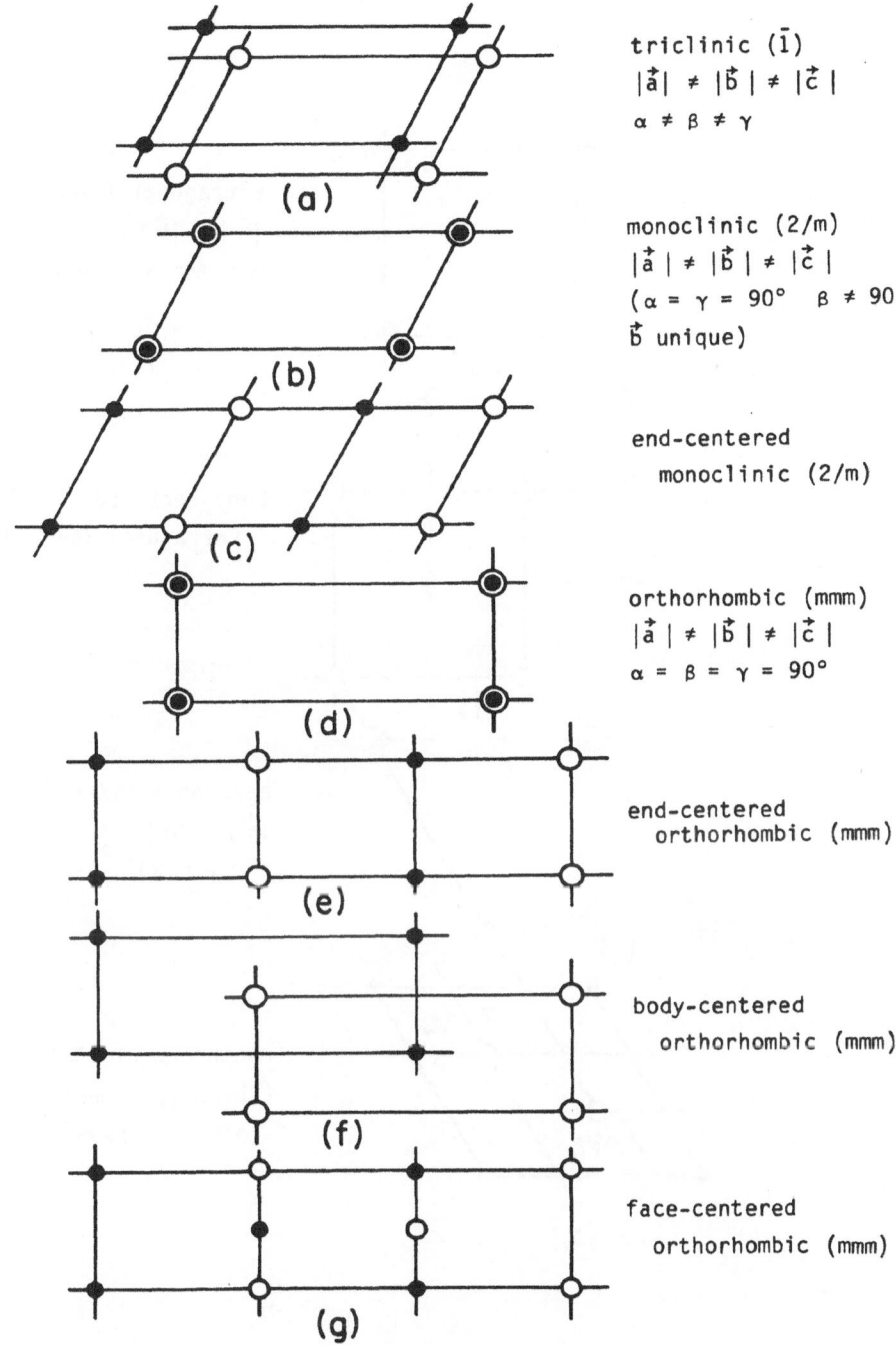

triclinic ($\bar{1}$)
$|\vec{a}| \neq |\vec{b}| \neq |\vec{c}|$
$\alpha \neq \beta \neq \gamma$

(a)

monoclinic (2/m)
$|\vec{a}| \neq |\vec{b}| \neq |\vec{c}|$
($\alpha = \gamma = 90°$ $\beta \neq 90°$
$\vec{b}$ unique)

(b)

end-centered
  monoclinic (2/m)

(c)

orthorhombic (mmm)
$|\vec{a}| \neq |\vec{b}| \neq |\vec{c}|$
$\alpha = \beta = \gamma = 90°$

(d)

end-centered
  orthorhombic (mmm)

(e)

body-centered
  orthorhombic (mmm)

(f)

face-centered
  orthorhombic (mmm)

(g)

Fig. 7 (continued)

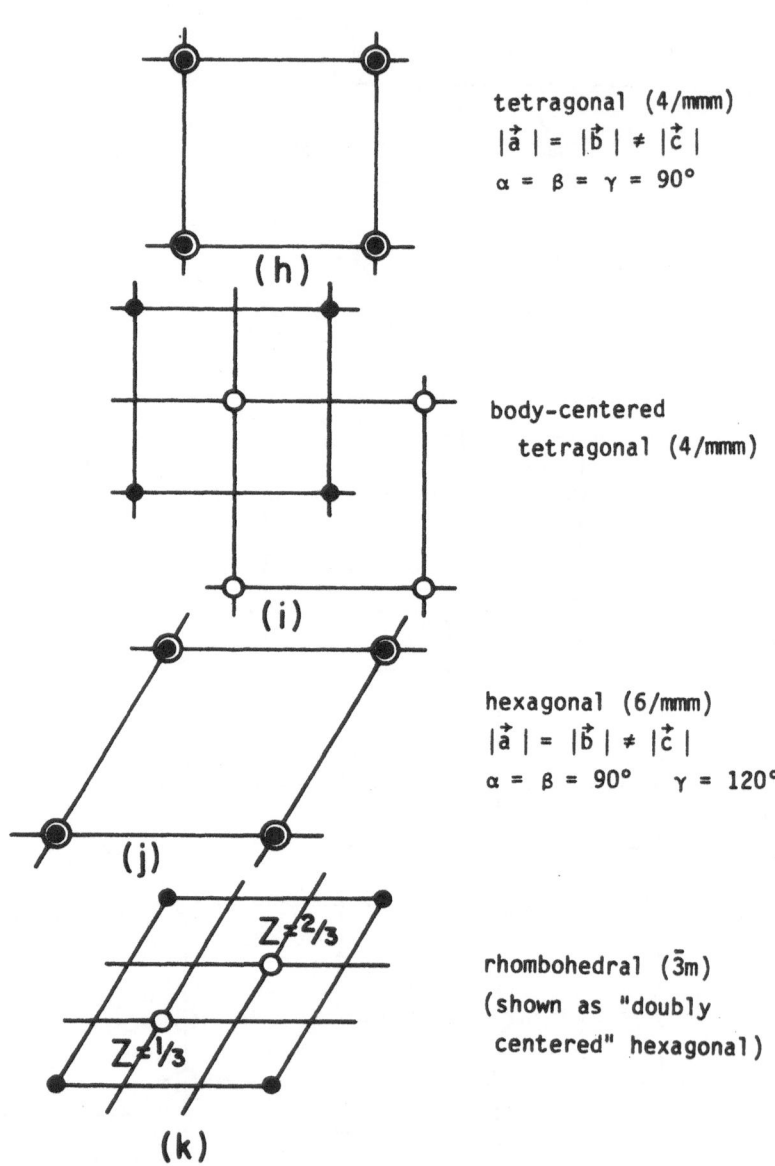

tetragonal (4/mmm)
$|\vec{a}| = |\vec{b}| \neq |\vec{c}|$
$\alpha = \beta = \gamma = 90°$

(h)

body-centered
tetragonal (4/mmm)

(i)

hexagonal (6/mmm)
$|\vec{a}| = |\vec{b}| \neq |\vec{c}|$
$\alpha = \beta = 90°$ $\gamma = 120°$

(j)

$Z = ^2/_3$

$Z = ^1/_3$

rhombohedral ($\bar{3}$m)
(shown as "doubly
centered" hexagonal)

(k)

Fig. 7 (continued)

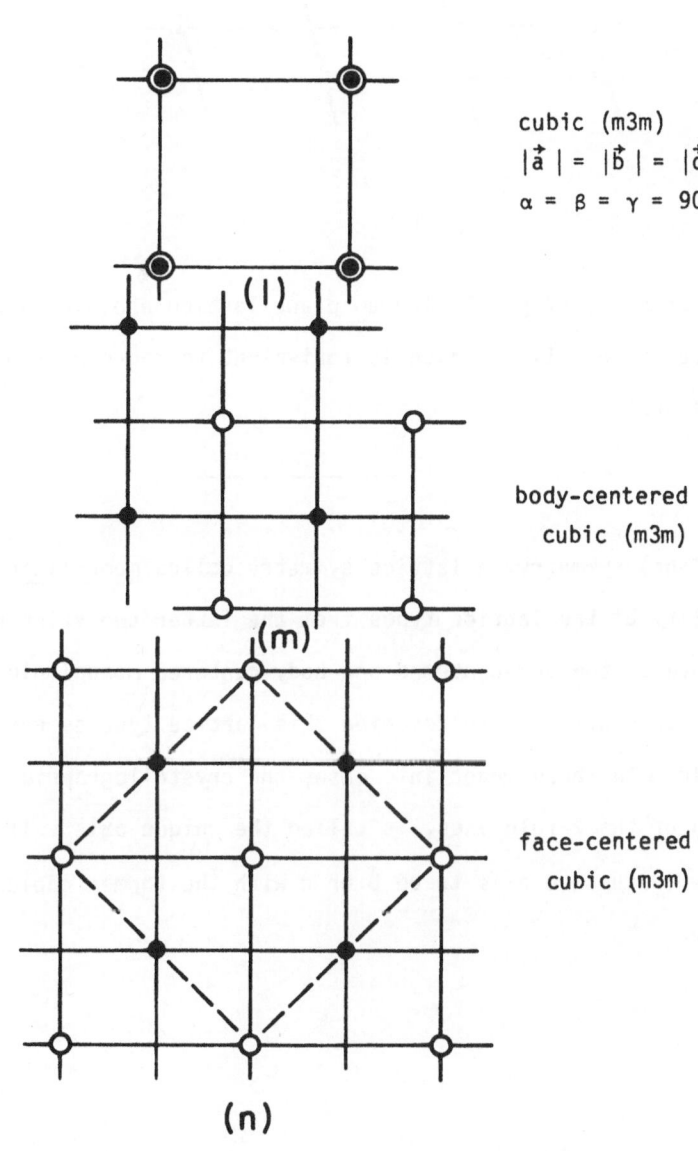

cubic (m3m)
$|\vec{a}| = |\vec{b}| = |\vec{c}|$
$\alpha = \beta = \gamma = 90°$

(l)

body-centered
cubic (m3m)

(m)

face-centered
cubic (m3m)

(n)

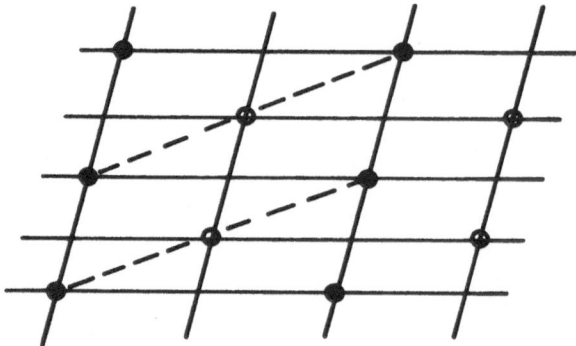

Fig. 8. Stacking of parallelogram plane lattice above base center yields
body-centered monoclinic which is equivalent to end-centered monoclinic
(dashed lines).

mirror plane) symmetry, a lattice symmetry called <u>monoclinic</u>. Fig. 8 shows

the equality of the lattice types from the latter two stackings, i.e., the

equivalence of the end-centered and body-centered monoclinic lattice types.

The usual convention is to describe this lattice type as end-centered

monoclinic. In these monoclinic cases the crystallographic axis along the

direction of the 2-fold axes, is called the unique axis. It is customary to

designate the unique axis to be $\vec{b}$ or $\vec{c}$ with the former choice generally

preferred.

With the $\vec{b}$ axis chosen as the unique axis the β angle (the angle
angle between $\vec{a}$ and $\vec{b}$) is the interaxial angle that differs from 90°. The
centered face, when one occurs, is named according to the crystallographic
axis not included in the face, and the usual monoclinic convention is to
take this as the C face, i.e., the face formed by $\vec{a}$ and $\vec{b}$. Thus the
rectangular face of the end-centered monoclinic cell is conventionally the
a-b face and the translations implied by C-centering are given by $\frac{\vec{a}+\vec{b}}{2}$ added
to all translational symmetry operations implied by $\vec{a}$, $\vec{b}$, and $\vec{c}$.

It is also possible to obtain monoclinic space lattices by stacking
rectangular plane lattices, primitive rectangular stacked with $\vec{c}$ projecting
onto one edge of the basal unit cell yields primitive monoclinic, and
centered rectangular similarly stacked yields an end-centered monoclinic
space lattice.

The rectangular lattices can also be stacked to yield space lattices
with 2/m 2/m 2/m (three mutually orthogonal mirror planes with three mutually
orthogonal two-fold axes along their intersection lines) symmetry in four
different ways. The three stacking modes shown in Fig. 7d-f result from the
stacking of primitive rectangular lattices (the space lattices of 7e and 7f
could also be produced by stacking centered rectangular lattices), and 7g
results from the stacking of centered rectangular lattices. In each case the
stacking is done so as to maintain the 2-fold and mirror symmetries of the
plane lattices in the resulting space lattices, and a horizontal mirror plane
and horizontal two-fold axes are found in the resulting orthorhombic space
lattices.

The plane square lattice type can be stacked in five different ways
that maintain the 4-fold rotational symmetry in the space lattices. One

places the lattice points of the stacked plane directly above those of the basal plane, at an arbitrary height (7h) and a second places the lattice points of the stacked plane above the centers of the basal unit cells at an arbitrary height (7i). In both of these cases the space lattice has 4/mmm (a four-fold axis with two vertical and one horizontal mirror planes)symmetry and the space lattice types are primitive and body-centered <u>tetragonal</u>, respectively. It is left as an exercise to show how a face-centered tetragonal lattice is generated and to demonstrate that the resultant lattice type is equivalent to a body-centered tetragonal space lattice. The tetragonal space lattices are also generated from rectangular plane lattices if a vertical stacking vector is the same length as either $\vec{a}$ or $\vec{b}$.

When the stacking heights of the square lattices are not arbitrary it is possible to introduce additional symmetry elements. When the vertical stacking of Fig. 7(ℓ) occurs with $|\vec{a}| = |\vec{b}| = |\vec{c}|$ a simple cubic lattice results. When the stacking of 7(m) occurs with the stacked layer at height $|\vec{a}|/2$ a body-centered cubic lattice results, and when the stacking height is $|\vec{a}|/\sqrt{2}$ a face-centered cubic lattice is obtained (Fig. 7(n)). Note that the resulting cubic symmetry includes 3-fold axes along the body diagonals, and that it therefore follows that cubic lattices could also be generated by the appropriate stacking of hexagonal plane lattices.

The remaining cases are those that result from stacking of plane lattices with 6-fold rotational symmetry. If $\vec{c}$ is perpendicular to $\vec{a}$ and $\vec{b}$ then primitive hexagonal space lattices result. If, on the other hand, the 6-fold axis of the stacked lattice is coincident with the 3-fold axis of the base lattice then the stacking vector must be repeated twice before a complete unit cell of a space lattice with 3-fold symmetry is obtained. The resulting hexagonal cell is triply primitive, and the primitive cell is rhombohedral, as shown in Fig. 9.

The 14 Bravais lattices are thus generated by the stacking of plane lattices. The fact that there exist no more than 14 space lattice types is made plausible by the limitations placed upon the rotational symmetries of plane lattices by the lattice periodicity and by the limitations placed on the number of ways that plane lattices can stack by the requirement that proper rotational axes superimpose if symmetry is to be maintained in the space lattices.

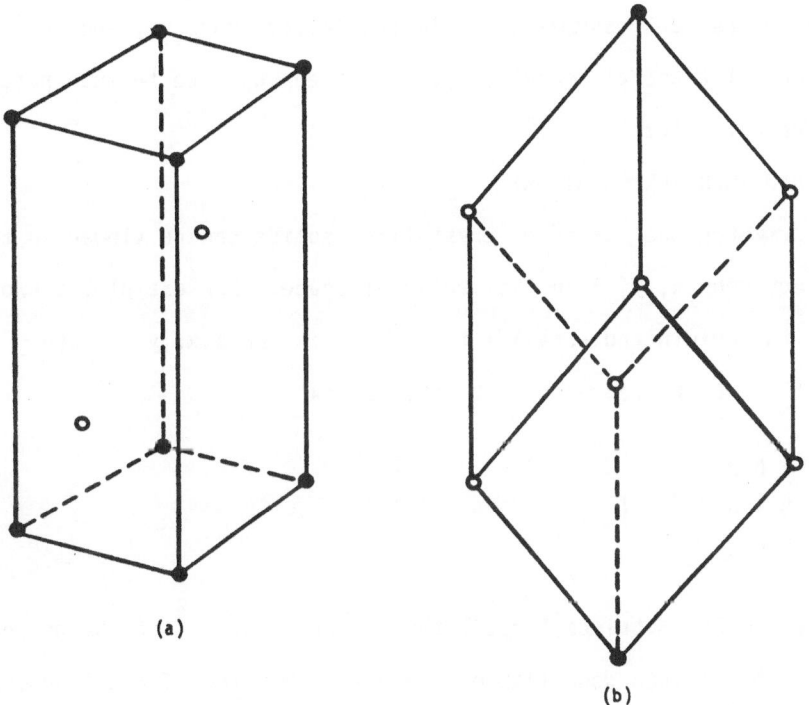

(a)

(b)

Fig. 9(a) Triply primitive hexagonal cell.

(b) Primitive rhombohedral cell.

SPACE GROUP SYMMETRY

I. INTRODUCTION

Thus far the treatment of symmetry has been restricted to the proper rotational and reflection symmetries of space lattices. The discussion of the symmetry of crystalline solids does not end with the presentation of the 14 Bravais lattice types because the symmetry of a solid is the symmetry of its three-dimensionally periodic particle density, and there are more symmetries available to such periodic patterns than to the lattices which characterize their translational symmetries. This is in part because the pattern need not be centrosymmetric while the lattice must be, and in part because of the existence of symmetry operations appropriate to such patterns but not to their lattices.

II. PROPER AND IMPROPER ROTATIONS

The symmetry operations of crystalline solids can be viewed in terms of their effect upon x, y, z in 3-dimensional space. For example, a two-fold axis through the origin and parallel to the $\vec{c}$ axis takes x, y, z into $\bar{z}$, $\bar{y}$, z and thus can be represented by the matrix

$$\begin{pmatrix} \bar{1} & 0 & 0 \\ 0 & \bar{1} & 0 \\ 0 & 0 & 1 \end{pmatrix} \begin{pmatrix} x \\ y \\ z \end{pmatrix} = \begin{pmatrix} \bar{x} \\ \bar{y} \\ z \end{pmatrix} \quad ,$$

and similarly the 3x3 matrices for all the proper rotations of the groups O_h and D_{6h} can be written down (Tables 1 and 2). Matrices for all proper rotations of all space groups are contained in these tables. Fig. 10 clarifies the perhaps not immediately obvious relationship of x, y, z and the point related to it by a C_{6z} operation (x-y, x, z) in a hexagonal lattice.

TABLE 1. PROPER ROTATIONAL SYMMETRY OPERATIONS OF O_h

$$\begin{pmatrix} 1 & 0 & 0 \\ 0 & 1 & 0 \\ 0 & 0 & 1 \end{pmatrix} \varepsilon \qquad\qquad \begin{pmatrix} 0 & \bar{1} & 0 \\ 0 & 0 & \bar{1} \\ 0 & 0 & 1 \end{pmatrix} C_{3(x-y+z)}$$

$$\begin{pmatrix} 1 & 0 & 0 \\ 0 & \bar{1} & 0 \\ 0 & 0 & \bar{1} \end{pmatrix} C_{2x} \qquad\qquad \begin{pmatrix} 0 & 0 & 1 \\ 1 & 0 & 0 \\ 0 & 1 & 0 \end{pmatrix} C_{3(x+y+z)}$$

$$\begin{pmatrix} \bar{1} & 0 & 0 \\ 0 & 1 & 0 \\ 0 & 0 & \bar{1} \end{pmatrix} C_{2y} \qquad\qquad \begin{pmatrix} 0 & 0 & 1 \\ \bar{1} & 0 & 0 \\ 0 & \bar{1} & 0 \end{pmatrix} C^2_{3(x-y+z)}$$

$$\begin{pmatrix} \bar{1} & 0 & 0 \\ 0 & \bar{1} & 0 \\ 0 & 0 & 1 \end{pmatrix} C_{2z} \qquad\qquad \begin{pmatrix} 0 & 0 & \bar{1} \\ 1 & 0 & 0 \\ 0 & \bar{1} & 0 \end{pmatrix} C^2_{3(x+y-z)}$$

$$\begin{pmatrix} 0 & 1 & 0 \\ 0 & 0 & 1 \\ 1 & 0 & 0 \end{pmatrix} C^2_{3(x+y+z)} \qquad\qquad \begin{pmatrix} 0 & 0 & 1 \\ 0 & 1 & 0 \\ \bar{1} & 0 & 0 \end{pmatrix} C_{4y}$$

$$\begin{pmatrix} 0 & 1 & 0 \\ 0 & 0 & \bar{1} \\ \bar{1} & 0 & 0 \end{pmatrix} C_{3(x+y-z)} \qquad\qquad \begin{pmatrix} 0 & 0 & \bar{1} \\ \bar{1} & 0 & 0 \\ 0 & 1 & 0 \end{pmatrix} C^2_{3(-x+y+z)}$$

$$\begin{pmatrix} 0 & \bar{1} & 0 \\ 0 & 0 & 1 \\ \bar{1} & 0 & 0 \end{pmatrix} C_{3(-x+y+z)} \qquad\qquad \begin{pmatrix} 0 & \bar{1} & 0 \\ \bar{1} & 0 & 0 \\ 0 & 0 & \bar{1} \end{pmatrix} C_{2(-x+y)}$$

TABLE 1 - continued

$$\begin{pmatrix} 0 & \bar{1} & 0 \\ 1 & 0 & 0 \\ 0 & 0 & 1 \end{pmatrix} C_{4z}$$

$$\begin{pmatrix} 0 & 0 & \bar{1} \\ 0 & \bar{1} & 0 \\ \bar{1} & 0 & 0 \end{pmatrix} C_{2(-y+z)}$$

$$\begin{pmatrix} 0 & 1 & 0 \\ \bar{1} & 0 & 0 \\ 0 & 0 & 1 \end{pmatrix} C_{4z}^3$$

$$\begin{pmatrix} 0 & 0 & \bar{1} \\ 0 & 1 & 0 \\ 1 & 0 & 0 \end{pmatrix} C_{4y}^3$$

$$\begin{pmatrix} 0 & 1 & 0 \\ 1 & 0 & 0 \\ 0 & 0 & \bar{1} \end{pmatrix} C_{2(x+y)}$$

$$\begin{pmatrix} 0 & 0 & 1 \\ 0 & \bar{1} & 0 \\ 1 & 0 & 0 \end{pmatrix} C_{2(x+z)}$$

$$\begin{pmatrix} \bar{1} & 0 & 0 \\ 0 & 0 & \bar{1} \\ 0 & \bar{1} & 0 \end{pmatrix} C_{2(-y+z)}$$

$$\begin{pmatrix} \bar{1} & 0 & 0 \\ 0 & 0 & 1 \\ 0 & 1 & 0 \end{pmatrix} C_{2(y+z)}$$

$$\begin{pmatrix} 1 & 0 & 0 \\ 0 & 0 & \bar{1} \\ 0 & 1 & 0 \end{pmatrix} C_{4x}$$

$$\begin{pmatrix} 1 & 0 & 0 \\ 0 & 0 & 1 \\ 0 & 1 & 0 \end{pmatrix} C_{4x}^3$$

TABLE 2. PROPER SYMMETRY OPERATIONS OF D_{6h}

$$\begin{pmatrix} 1 & 0 & 0 \\ 0 & 1 & 0 \\ 0 & 0 & 1 \end{pmatrix}\ \varepsilon \qquad\qquad \begin{pmatrix} \bar{1} & 0 & 0 \\ \bar{1} & 1 & 0 \\ 0 & 0 & 1 \end{pmatrix}\ C_{2y}$$

$$\begin{pmatrix} 1 & \bar{1} & 0 \\ 1 & 0 & 0 \\ 0 & 0 & 1 \end{pmatrix}\ C_{6z} \qquad\qquad \begin{pmatrix} 0 & \bar{1} & 0 \\ \bar{1} & 0 & 0 \\ 0 & 0 & \bar{1} \end{pmatrix}\ C_{2(x-y)}$$

$$\begin{pmatrix} 0 & \bar{1} & 0 \\ 1 & \bar{1} & 0 \\ 0 & 0 & 1 \end{pmatrix}\ C_{3z} \qquad\qquad \begin{pmatrix} 1 & \bar{1} & 0 \\ 0 & \bar{1} & 0 \\ 0 & 0 & \bar{1} \end{pmatrix}\ C_{2x}$$

$$\begin{pmatrix} \bar{1} & 0 & 0 \\ 0 & \bar{1} & 0 \\ 0 & 0 & 1 \end{pmatrix}\ C_{2z} \qquad\qquad \begin{pmatrix} 1 & 0 & 0 \\ 1 & \bar{1} & 0 \\ 0 & 0 & \bar{1} \end{pmatrix}\ C_{2(2x+y)}$$

$$\begin{pmatrix} \bar{1} & 1 & 0 \\ \bar{1} & 0 & 0 \\ 0 & 0 & 1 \end{pmatrix}\ C_{3z}^{2} \qquad\qquad \begin{pmatrix} \bar{1} & 1 & 0 \\ 0 & 1 & 0 \\ 0 & 0 & \bar{1} \end{pmatrix}\ C_{2(x+2y)}$$

$$\begin{pmatrix} 0 & 1 & 0 \\ \bar{1} & 1 & 0 \\ 0 & 0 & 1 \end{pmatrix}\ C_{6z}^{5}$$

The proper rotational symmetry operations of a crystalline solid form a closed set (in fact a group). This implies that if two such operations are members of the set then the combined operation is also. For example, in O_h the 90° rotation along z together with the 120° rotation along the body diagonal $(C_{3(x+y+z)})$ combine according to

$$\begin{pmatrix} 0 & \bar{1} & 0 \\ 1 & 0 & 0 \\ 0 & 0 & 1 \end{pmatrix} \begin{pmatrix} 0 & 0 & 1 \\ 1 & 0 & 0 \\ 0 & 1 & 0 \end{pmatrix} = \begin{pmatrix} \bar{1} & 0 & 0 \\ 0 & 0 & 1 \\ 0 & 1 & 0 \end{pmatrix}$$

and, by inspection of Table 1, the resultant matrix represents $C_{2(y+z)}$, the 180° rotation about a face diagonal.

In the space groups, if all symmetries are to be considered, it is necessary to include operations which are represented by the matrices of Tables 1 and 2 with the signs of all nonzero matrix elements changed. The resultant matrices represent operations of rotation followed by inversion (not necessarily the addition of inversion as a symmetry element) and are called rotoinversion operations. Figures which portray the symmetries generated by the resulting rotoinversion axes ($\bar{1},\bar{2},\bar{3},\bar{4}$ and $\bar{6}$, for, as will be shown below, rotoinversion (improper) axes are compatible with translational periodicity only if they are of the same order as the allowed proper axes) are shown in Fig. 11. Note that while the odd improper axes imply a center of symmetry, the even improper axes do not. Considering symmetry elements, the $\bar{1}$ axis is equivalent to an inversion center, the $\bar{2}$ axis to a mirror plane, the $\bar{3}$ axis to a 3-fold proper axis and a center of inversion and the $\bar{6}$ axis to a 3-fold axis and a horizontal mirror plane. The improper 4-fold axis is not equivalent to any other element or combination of symmetry elements.

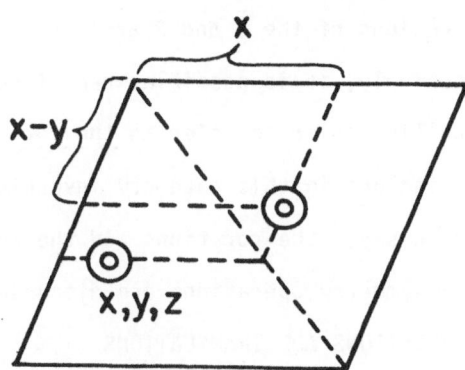

Fig. 10. Illustration of relationship of positional parameters under C_{6z} in a hexagonal lattice.

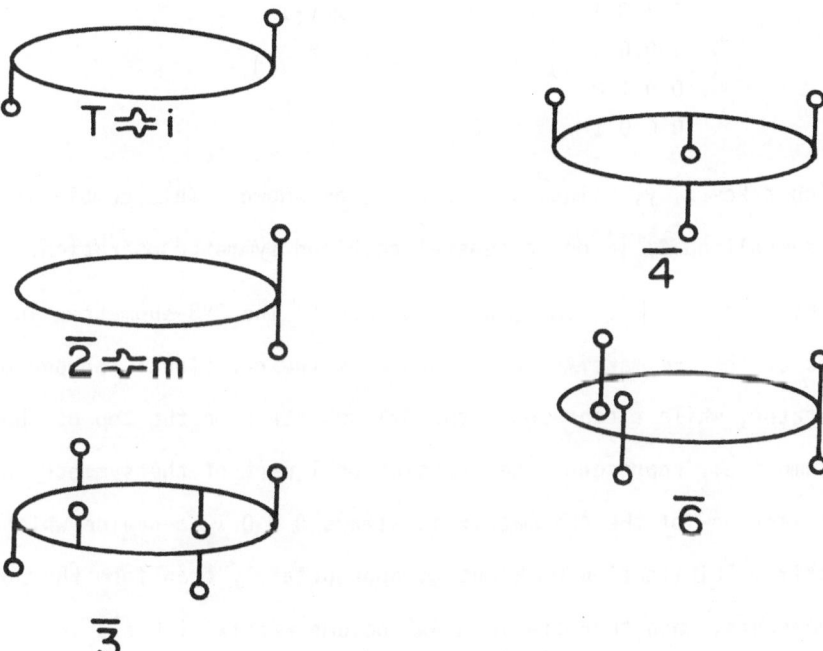

Fig. 11. Symmetries of the improper axes.

Since the operations of the $\bar{1}$ and $\bar{2}$ axes are operations of inversion and reflection, respectively, it is possible when discussing the symmetry operations of a crystalline solid to refer to <u>the rotations</u> (proper and improper) and thereby include in this category inversions and reflections, making it unnecessary to say, "the rotations and the reflections and the inversions" when these symmetry operations are discussed.

III. COMBINATION OF ROTATIONS AND TRANSLATIONS

As was done earlier it is possible to represent combined rotational and translational symmetry operations in a single matrix, for example the C_{6z} operation and translation by $\vec{a}$ in D_{6h} is represented by the 4x4 matrix (Seitz operator)

$$\begin{pmatrix} 1 & \bar{1} & 0 & 1 \\ 1 & 0 & 0 & 0 \\ 0 & 0 & 1 & 0 \\ 0 & 0 & 0 & 1 \end{pmatrix} \begin{pmatrix} x \\ y \\ z \\ 1 \end{pmatrix} = \begin{pmatrix} x-y+1 \\ x \\ z \\ 1 \end{pmatrix}$$

which takes x, y, z into x-y+1, x, z, as shown. This combined operation can be symbolized $C_{6z} | \vec{a}$ or, a general combined symmetry operation, can be symbolized by $\beta | \vec{t}$. The β part represents the 3x3 submatrix in the upper left of the 4x4 matrix i.e., represents the rotational (proper or improper) operator, while $\vec{t}$ represents the 3x1 submatrix on the top of the last column i.e., represents the translational part of the symmetry operation. The last row of the 4x4 matrix is always 0 0 0 1, a device which makes the matrix multiplication work out to appropriately transform the positional parameters, when they are in a 4x1 column matrix with a 1 in the bottom row.

The symbol $\vec{t}$ was intentionally chosen (as opposed to $\vec{T}$) because in space group symmetries $\vec{t}$ need not be a lattice translation. This occurs in particular when the operation is a screw or glide operation, as discussed below.

IV. SCREW AXES AND GLIDE PLANES

First consider how the operations $\beta \mid \vec{t}$ combine. Consideration of the multiplication of the 4x4 matrices demonstrates that the β parts combine just as do the 3x3 matrices representing the rotations in the absence of a translational component (Tables 1 and 2), and the translation part of the operation performed first (that to the right) is rotated by the rotation of that performed second (that to the left) and combined with the translational component of that performed second, i.e.,

$$\beta_2 \mid \vec{t}_2 \cdot \beta_1 \mid \vec{t}_1 = \beta_2\beta_1 \mid \beta_2\vec{t}_1 + \vec{t}_2.$$

If $\beta \mid \vec{t}$ is a combined rotation-translation operation, then operation n times with this operation, where $\beta^n = \epsilon$, yields

$$\epsilon \mid \beta^n \vec{t} + \beta^{n-1} \vec{t} + \ldots + \vec{t}$$

which must be a pure translation if $\beta \mid \vec{t}$ is to be a possible symmetry operation. For example consider $C_{2z} \mid \vec{c}/2$, where

$$C_{2z} \mid \vec{c}/2 \cdot C_{2z} \mid \vec{c}/2 = \epsilon \mid \vec{c}.$$

Since $\vec{c}$ is a pure translation it follows that $C_{2z} \mid \vec{c}/2$ is a possible symmetry operation. This operation is one of a 2_1 screw axis as shown in Fig. 12. This operation is represented by

$$\begin{pmatrix} \bar{1} & 0 & 0 & 0 \\ 0 & \bar{1} & 0 & 0 \\ 0 & 0 & 1 & 1/2 \\ 0 & 0 & 0 & 1 \end{pmatrix}$$

which takes x, y, z into $\bar{x}$, $\bar{y}$, $z+1/2$ (the axis has been taken to pass through x=y=0). The product of the matrix with itself,

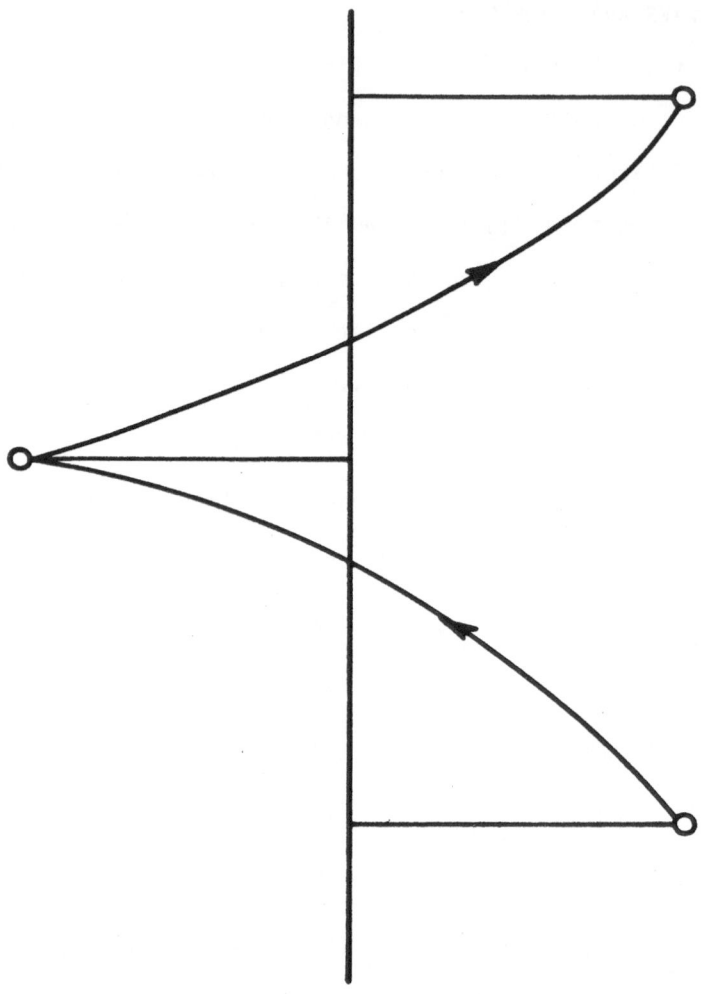

Fig. 12. The operation of a 2_1 axis.

$$
\begin{pmatrix}
\bar{1} & 0 & 0 & 0 \\
0 & \bar{1} & 0 & 0 \\
0 & 0 & 1 & {}^{1}/_{2} \\
0 & 0 & 0 & 1
\end{pmatrix}
\begin{pmatrix}
\bar{1} & 0 & 0 & 0 \\
0 & \bar{1} & 0 & 0 \\
0 & 0 & 1 & {}^{1}/_{2} \\
0 & 0 & 0 & 1
\end{pmatrix}
=
\begin{pmatrix}
1 & 0 & 0 & 0 \\
0 & 1 & 0 & 0 \\
0 & 0 & 1 & 1 \\
0 & 0 & 0 & 1
\end{pmatrix}
,
$$

in agreement with the above, represents $\vec{c}$, a pure translation. Note that such rotational operations of a space group containing a 2_1 axis do not form a subgroup of the space group since the rotational space group operations are not closed under binary combination.

Applying $\beta^n = \epsilon \rightarrow \beta^n \vec{t} + \beta^{n-1} \vec{t} + \ldots + \vec{t} = \vec{T}$ to $\beta = C_2, C_3, C_4$ and C_6 (which will be shown to yield all of the possible screw axes) yields the axes summarized in Table 3. The <u>rotational symmetry operations</u> can be defined as those operations for which the β part is not ϵ, and the remaining operations are the pure translations $\{\epsilon \mid \vec{T}_{mnp}\}$. The rotational symmetry operations in space groups can then be divided into <u>essential</u> and <u>implied</u> operations, where the essential operations are those which contain no pure translational part (no $\vec{T}_{mnp}$) and the implied operations are those which result from essential operations by combination with pure translations. In general there are some essential and/or implied axes (proper, improper, screw) which do not pass through a lattice point. This point was discussed for proper axes and space lattices in a previous section. A general screw-rotational operation, $\beta \mid \vec{t} + \vec{t}'$ has a translation component along the axis about which β operates, ($\vec{t}$), and a component perpendicular to the axis ($\vec{t}'$). If $\vec{e}$ is a vector from the origin to the axis and perpendicular to the axis, then the total operation can be considered to be performed by: 1. translation by $-\vec{e}$, 2. rotation by β about the origin, 3. translation by $\vec{e}$, 4. translation by $\vec{t}$, i.e.,

$$2_1$$
$$3_1 \qquad 3_2$$
$$4_1 \qquad 4_2 \qquad 4_3$$
$$6_1 \qquad 6_2 \qquad 6_3 \qquad 6_4 \qquad 6_5$$

TABLE 3. The allowed screw axes Nm, where N represents the rotation and m/N the fractional translation along the axis.

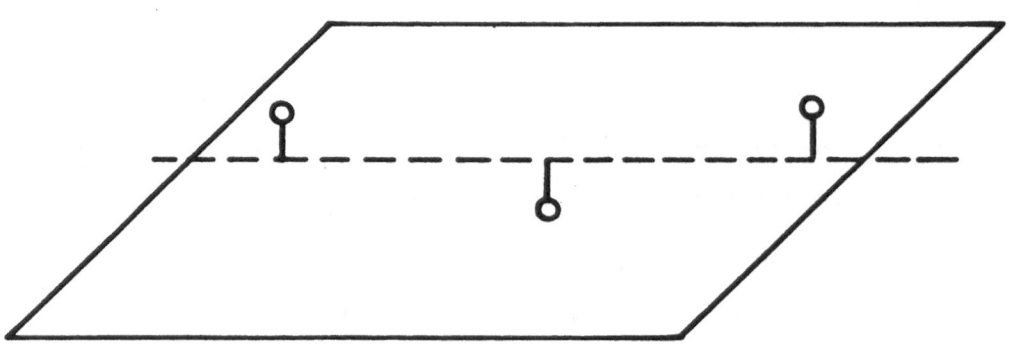

Fig. 13. The glide symmetry operations.

$$\beta \,|\, \vec{t} + \vec{t}' = \epsilon \,|\, \vec{e} \cdot \beta \,|\, 0 \cdot \epsilon \,|\, -\vec{e} + \epsilon \,|\, \vec{t}$$

$$= \beta \,|\, \beta(-\vec{e}) + \vec{e} + \vec{t},$$

and

$$\vec{t}' = \beta(-\vec{e}) + \vec{e} .$$

Thus the existence in $\beta|\vec{t} + \vec{t}'$ of a component of translation perpendicular to the axis implies that the rotation is about an axis which does not pass through the origin. For example

$$\begin{pmatrix} 1 & 0 & 0 & 1/2 \\ 0 & \bar{1} & 0 & 1/2 \\ 0 & 0 & \bar{1} & 1/2 \\ 0 & 0 & 0 & 1 \end{pmatrix}$$

which can be written $C_{2x}\,|\, \dfrac{\vec{a} + \vec{b} + \vec{c}}{2}$, is an operation of a 2_1 axis along x through a point given by

$$\frac{\vec{b} + \vec{c}}{2} = C_{2x} \,(-\vec{e}) + \vec{e} .$$

Since $\vec{e}$ is perpendicular to the axis, $C_{2x}(-\vec{e}) = \vec{e}$ and thus $\vec{e} = \dfrac{\vec{b}+\vec{c}}{4}$ i.e. the

2_1 axis for which the above matrix represents an operation passes through the unit cell at the points with $y = z = 1/4$.

Another kind of combination of a β operation with a non-lattice translation that occurs as a symmetry operation in space groups is that of reflection combined with translation parallel to the plane of reflection (Fig. 13). Since reflection is its own inverse the translation $2\vec{t}$ parallel to the plane must be a translational symmetry operation. The glide operations that result are labeled according to the direction of the glide, e.g., an $\vec{a}$ glide has translation component $(\vec{a}/2)$ parallel to the reflection plane, and an n-glide is a diagonal glide across a face.

It is possible that a mirror or glide plane does not pass through the origin of the unit cell, for example while

$$\begin{pmatrix} \bar{1} & 0 & 0 & 0 \\ 0 & 1 & 0 & 1/2 \\ 0 & 0 & 1 & 1/2 \\ 0 & 0 & 0 & 1 \end{pmatrix}$$

represents an n-glide perpendicular to $\vec{a}$ with the reflection plane containing the origin,

$$\begin{pmatrix} \bar{1} & 0 & 0 & 1/2 \\ 0 & 1 & 0 & 1/2 \\ 0 & 0 & 1 & 1/2 \\ 0 & 0 & 0 & 1 \end{pmatrix}$$

represents the same operation with the plane of reflection at $x = 1/4$. The location of the plane of reflection is determined by the component of the translation perpendicular to the reflection plane. In this case

and
$$\vec{a}/2 = \sigma_x (-\vec{e}) + \vec{e}$$
$$\vec{e} = 1/4 \, \vec{a} \quad \text{(see Fig. 14).}$$

V. COMBINATION OF SYMMETRY OPERATIONS

The rotational symmetry of a lattice that is consistent with a space group with only the rotational symmetry operations of a 2_1 axis is 2, with only those of a 3_1 axis is 3, with those of a 6_3 axis is 6, and so forth. That is, the rotational symmetries of a lattice corresponding to a given space group are the β parts of the $\beta | \vec{t}$ operations of the space group. To prove this is so, consider a space group in which $\beta | \vec{t}$ and $\epsilon | \vec{\tau}$ are symmetry operations. The product

$$\beta | \vec{t} \cdot \epsilon | \vec{\tau} \cdot \beta^{-1} | -\beta^{-1} \vec{t} = \epsilon | \beta\vec{\tau}$$

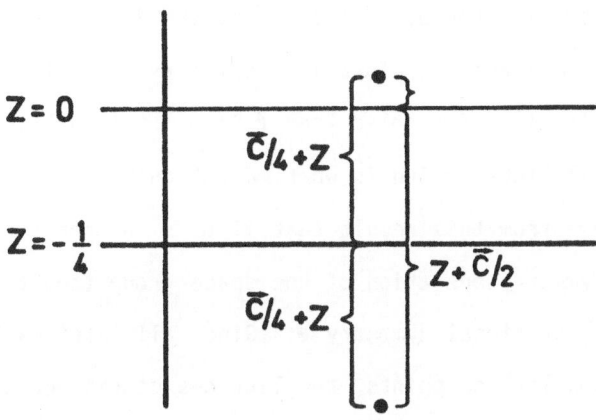

Fig. 14. Reflection through z=0 followed by translation by $\vec{c}/2$ is equivalent to reflection through $\vec{c}/4$.

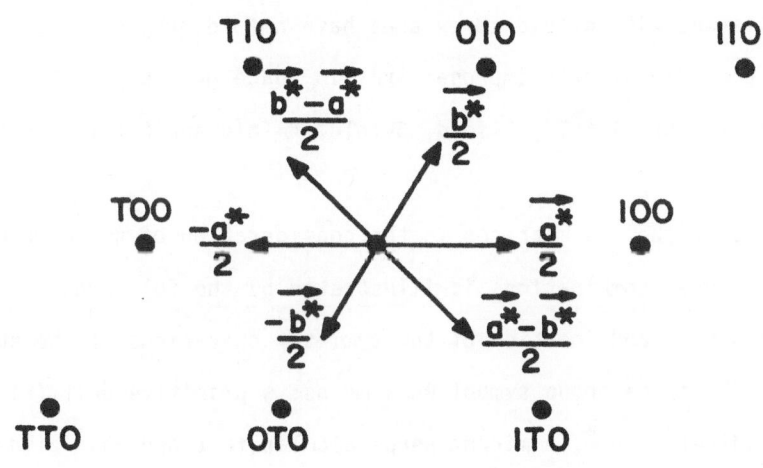

Fig. 15. Hexagonal reciprocal lattice showing $\vec{a}^*/2$ and all the $\vec{k}$ vectors that result from $\vec{a}^*/2$ by symmetry operations of D_{6h}.

must be a symmetry operation because $\beta^{-1}|\beta^{-1}\, \vec{t}$ is the inverse of $\beta|\vec{t}$
$(\beta^{-1}|-\beta^{-1}\, \vec{t} \cdot \beta|\vec{t} = \epsilon|0)$ and therefore is a symmetry operation. Thus if
$\beta|\vec{t}$ is a symmetry operation of a solid then β is a rotational symmetry
operation of the translations, which is what we set out to prove.

It also follows from this result that if β is an operator for an
improper rotational symmetry operation of the space group then the lattice
also has the improper rotational symmetry β. Since all lattices are
centrosymmetric through lattice points, the lattices consistent with solids
with improper axes will have at least the symmetry required by the addition
of an inversion center to the improper axis, i.e., $\bar{2}$ in the structure implies
2-fold lattice symmetry, $\bar{3}$ in the structure implies 3-fold lattice symmetry,
$\bar{4}$ in the structure implies 4-fold lattice symmetry and $\bar{6}$ in the structure
implies 6-fold lattice symmetry. Thus the lattices consistent with
structures with n-fold rotoinversion axes have n-fold proper axes, and the
lattices consistent with n-fold screw axes have n-fold proper axes. It
follows that the only allowed improper axes in space groups and the only
allowed screw axes are 1-fold, 2-fold, 3-fold, 4-fold and 6-fold, as stated
previously.

The use of the 4x4 matrices in the consideration of space group
operations and their combinations is illustrated by the following problem,
"what are the nature and location of the symmetry operations of the space
group Pnma"? The space group symbol Pnma means: a primitive cell (P) with an
n-glide perpendicular to $\vec{a}$, a mirror perpendicular to $\vec{b}$ and an $\vec{a}$ glide
perpendicular to $\vec{c}$. Thus the β (rotational) parts of the symmetry operations
are σ_x, σ_y and σ_z from which it follows that the crystal class is D_{2h}.
Since it is not clear at the outset where to place the glides and mirror
relative to the origin, we take the translational components perpendicular to

the reflection planes to be unspecified: $\vec{t}_1$, $\vec{t}_2$ and $\vec{t}_3$. Then the n-glide perpendicular to $\vec{a}$ is represented by

$$\begin{pmatrix} \bar{1} & 0 & 0 & t_1 \\ 0 & 1 & 0 & 1/2 \\ 0 & 0 & 1 & 1/2 \\ 0 & 0 & 0 & 1 \end{pmatrix}$$

the mirror perpendicular to $\vec{b}$ by

$$\begin{pmatrix} 1 & 0 & 0 & 0 \\ 0 & \bar{1} & 0 & t_2 \\ 0 & 0 & 1 & 0 \\ 0 & 0 & 0 & 1 \end{pmatrix}$$

and the a-glide perpendicular to $\vec{c}$ by

$$\begin{pmatrix} 1 & 0 & 0 & 1/2 \\ 0 & 1 & 0 & 0 \\ 0 & 0 & \bar{1} & t_3 \\ 0 & 0 & 0 & 1 \end{pmatrix} .$$

Three two-fold-type (i.e., two-fold or two-fold screw) operations are generated by pairwise combination of these reflection-type operations:

$$\begin{pmatrix} \bar{1} & 0 & 0 & t_1 \\ 0 & 1 & 0 & 1/2 \\ 0 & 0 & 1 & 1/2 \\ 0 & 0 & 0 & 1 \end{pmatrix} \begin{pmatrix} 1 & 0 & 0 & 0 \\ 0 & \bar{1} & 0 & t_2 \\ 0 & 0 & 1 & 0 \\ 0 & 0 & 0 & 1 \end{pmatrix} = \begin{pmatrix} \bar{1} & 0 & 0 & t_1 \\ 0 & \bar{1} & 0 & t_2+1/2 \\ 0 & 0 & 1 & 1/2 \\ 0 & 0 & 0 & 1 \end{pmatrix} ,$$

$$\begin{pmatrix} \bar{1} & 0 & 0 & t_1 \\ 0 & 1 & 0 & 1/2 \\ 0 & 0 & 1 & 1/2 \\ 0 & 0 & 0 & 1 \end{pmatrix} \begin{pmatrix} 1 & 0 & 0 & 1/2 \\ 0 & 1 & 0 & 0 \\ 0 & 0 & \bar{1} & t_3 \\ 0 & 0 & 0 & 1 \end{pmatrix} = \begin{pmatrix} \bar{1} & 0 & 0 & t_1-1/2 \\ 0 & 1 & 0 & 1/2 \\ 0 & 0 & \bar{1} & t_3+1/2 \\ 0 & 0 & 0 & 1 \end{pmatrix},$$

$$\begin{pmatrix} 1 & 0 & 0 & 0 \\ 0 & \bar{1} & 0 & t_2 \\ 0 & 0 & 1 & 0 \\ 0 & 0 & 0 & 1 \end{pmatrix} \begin{pmatrix} 1 & 0 & 0 & 1/2 \\ 0 & 1 & 0 & 0 \\ 0 & 0 & \bar{1} & t_3 \\ 0 & 0 & 0 & 1 \end{pmatrix} \begin{pmatrix} 1 & 0 & 0 & 1/2 \\ 0 & \bar{1} & 0 & t_2 \\ 0 & 0 & \bar{1} & t_3 \\ 0 & 0 & 0 & 1 \end{pmatrix} .$$

Next the combination of any one of the two-fold-type operations with the perpendicular reflection-type operation will yield an inversion:

$$\begin{pmatrix} \bar{1} & 0 & 0 & t_1 \\ 0 & \bar{1} & 0 & t_2-1/2 \\ 0 & 0 & 1 & 1/2 \\ 0 & 0 & 0 & 1 \end{pmatrix} \begin{pmatrix} 1 & 0 & 0 & 1/2 \\ 0 & 1 & 0 & 0 \\ 0 & 0 & \bar{1} & t_3 \\ 0 & 0 & 0 & 1 \end{pmatrix} = \begin{pmatrix} \bar{1} & 0 & 0 & t_1-1/2 \\ 0 & \bar{1} & 0 & t_2+1/2 \\ 0 & 0 & \bar{1} & t_3+1/2 \\ 0 & 0 & 0 & 1 \end{pmatrix} .$$

The location of this inversion center in the unit cell is fixed by fixing t_1, t_2, t_3. If the conventional choice of origin at the inversion center is made then $t_1 = 1/2$, $t_2 = -1/2$ and $t_3 = -1/2$ or, since $+1/2$ and $-1/2$ are equivalent as regards location of symmetry operations we can take $t_1 = t_2 = t_3 = 1/2$. Substituting this result into the matrices for the essential symmetry operations yields the following matrices (the corresponding essential symmetry elements are also given):

$$\begin{pmatrix} \bar{1} & 0 & 0 & 1/2 \\ 0 & 1 & 0 & 1/2 \\ 0 & 0 & 1 & 1/2 \\ 0 & 0 & 0 & 1 \end{pmatrix} \longrightarrow \text{an n-glide} \perp \text{to } \vec{a} \text{ at } \vec{a}/4 ,$$

$$\begin{pmatrix} 1 & 0 & 0 & 0 \\ 0 & \bar{1} & 0 & 1/2 \\ 0 & 0 & 1 & 0 \\ 0 & 0 & 0 & 1 \end{pmatrix} \longrightarrow \text{a mirror} \perp \text{to } \vec{b} \text{ at } \vec{b}/4 ,$$

$$
\begin{pmatrix}
1 & 0 & 0 & 1/2 \\
0 & 1 & 0 & 0 \\
0 & 0 & \bar{1} & 1/2 \\
0 & 0 & 0 & 1
\end{pmatrix}
$$
--→ an a-glide ⊥ to $\vec{c}$ at $\vec{c}/4$,

$$
\begin{pmatrix}
\bar{1} & 0 & 0 & 1/2 \\
0 & \bar{1} & 0 & 0 \\
0 & 0 & 1 & 1/2 \\
0 & 0 & 0 & 1
\end{pmatrix}
$$
--→ a 2_1 axis || to $\vec{c}$ at $x = 1/4$, $y = 0$,

$$
\begin{pmatrix}
\bar{1} & 0 & 0 & 0 \\
0 & 1 & 0 & 1/2 \\
0 & 0 & \bar{1} & 0 \\
0 & 0 & 0 & 1
\end{pmatrix}
$$
--→ a 2_1 axis || to $\vec{b}$ through $x = z = 0$,

$$
\begin{pmatrix}
1 & 0 & 0 & 1/2 \\
0 & \bar{1} & 0 & 1/2 \\
0 & 0 & \bar{1} & 1/2 \\
0 & 0 & 0 & 1
\end{pmatrix}
$$
--→ a 2_1 axis || to $\vec{a}$ through $y = z = 1/4$,

$$
\begin{pmatrix}
\bar{1} & 0 & 0 & 0 \\
0 & \bar{1} & 0 & 0 \\
0 & 0 & \bar{1} & 0 \\
0 & 0 & 0 & 1
\end{pmatrix}
$$
--→ an inversion center at the origin.

Other, implied symmetry operations in the unit cell can be generated by the combination of those given above with pure translations, for example the 2_1 axis parallel to $\vec{c}$ at $x = 1/4$, y=0 combines with translation by $\vec{b}$ to yield a 2_1 axis parallel to $\vec{c}$ at $x = 1/4$, $y = 1/2$, etc.

In the chapters that follow the group properties of space groups become important. It is therefore important to bear in mind that the symmetry operations of crystalline solids combine associatively, are closed under "multiplication", and include inverses for all operations and the identity operation i.e., the set $\{\beta_i \mid \vec{t}_i\}$ is a group. The set of all β_i's, $\{\beta_i\}$, is a point group called the crystal class. To each of the 32 crystal classes there corresponds a number of space groups that differ from each other in the sense that the essential axes may or may not be screw axes and the essential reflection planes may or may not be glide planes. For example the crystal class C_{2h} (ε, C_{2y}, σ_y and i) is consistent with a monoclinic lattice. To this crystal class belong the primitive space groups: P2/m, P2$_1$/m, P2/c, and P2$_1$/c. Among these only P2/m does not include screw axes or glide planes among its essential operations. Such a space group is called symmorphic. The other, nonsymmorphic, space groups can be considered to result from the systematic, allowed replacement of the pure rotational operations of P2/m by rotation-translation operations. The essential operations of the four groups with the origin taken to be at the inversion center are given in Table 4.

TABLE 4.

Class C_{2h}	C_{2y}	i	σ_y	ε
P2/m	$C_{2y} \mid 0$	$i \mid 0$	$\sigma_y \mid 0$	$\varepsilon \mid 0$
P2$_1$/m	$C_{2y} \mid \dfrac{\vec{b}}{2}$	$i \mid 0$	$\sigma_y \mid \dfrac{\vec{b}}{2}$	$\varepsilon \mid 0$
P2/c	$C_{2y} \mid \dfrac{\vec{c}}{2}$	$i \mid 0$	$\sigma_y \mid \dfrac{\vec{c}}{2}$	$\varepsilon \mid 0$
P2$_1$/c	$C_{2y} \mid \dfrac{\vec{b}+\vec{c}}{2}$	$i \mid 0$	$\sigma_y \mid \dfrac{\vec{b}+\vec{c}}{2}$	$\varepsilon \mid 0$

RECIPROCAL SPACE AND IRREDUCIBLE REPRESENTATIONS OF SPACE GROUPS

I. INTRODUCTION

The set of pure translational symmetry operations $\{\varepsilon \mid \vec{t}_i\}$ is a subgroup of the space group of a three dimensional crystalline solid. It is therefore meaningful to seek irreducible representations (irr. reps) and basis functions for this pure translational subgroup, and these play an important role in the theory of crystalline solids. For the benefit of the reader unfamiliar with group theory, a set of basis functions for a representation is made up of functions which transform into each other, or linear combinations thereof, under symmetry operations of the group. The irr. rep. is then the set of matrices which transform the functions under the symmetry operations (a representation) when the set cannot be reduced in the sense that a coordinate transformation results in splitting the basis functions into two or more sets of new functions each of which is a set of basis functions for a representation.

There is a theorem of group theory that says that if all of the operations of a group commute then the irr. reps. of the group are one dimensional i.e., the sets of basis functions contain only one function each and this function transforms into itself multiplied by a constant under the symmetry operations of the group. The pure translations do commute (vector addition is commutative) and thus the search for basis functions (and irr. reps.) of the pure translational subgroup is the search for functions which change by a multiplicative constant under a translational symmetry operation.

II. RECIPROCAL LATTICE

For example, we first seek a function which transforms into itself under the translational symmetry operations i.e., a function which has the periodicity of the lattice. This function is a basis function for the "totally symmetric" representation and its representation is a string of 1's, one for each $\vec{t}_i$.

For the purpose of finding a basis function for the totally symmetric representation of the pure translations it is useful to know a function of $\vec{r}$ that is incremented by an integer when $\vec{r}$ is incremented by $\vec{t}$. That is, it would help to have available a vector $\vec{K}$ such that

$$\vec{K} \cdot (\vec{r} + \vec{t}) = \vec{K} \cdot \vec{r} + \vec{K} \cdot \vec{t}$$

$$= \vec{K} \cdot \vec{r} + \text{integer},$$

for with such a vector, for example

$$\sin 2\pi \vec{K} \cdot (\vec{r} + \vec{t}) = \sin 2\pi [\vec{K} \cdot \vec{r} + \vec{K} \cdot \vec{t}]$$

$$= \sin 2\pi \vec{K} \cdot \vec{r}$$

and we have found, as desired, a function with the periodicity of the lattice. Thus we seek a $\vec{K}$ such that $\vec{K} \cdot \vec{t} = \text{integer}$. We know that $\vec{t} = m\vec{a} + n\vec{b} + p\vec{c}$ and therefore $\vec{K} = m^* \vec{a}^* + n^* \vec{b}^* + p^* \vec{c}^*$ with m^*, n^*, p^* integers and $\vec{a}^* \cdot \vec{a} = 1$, $\vec{a}^* \cdot \vec{b} = \vec{a}^* \cdot \vec{c} = 0$, $\vec{b}^* \cdot \vec{b} = 1$, $\vec{b}^* \cdot \vec{a} = \vec{b}^* \cdot \vec{c} = 0$, and $\vec{c}^* \cdot \vec{c} = 1$, $\vec{c}^* \cdot \vec{b} = \vec{c}^* \cdot \vec{a} = 0$ has the desired property since, with these definitions,

$$\vec{K} \cdot \vec{t} = m^* m + n^* n + p^* p$$

which is an integer. The definitions mean that $\vec{a}^*$ is perpendicular to $\vec{b}$ and $\vec{c}$, i.e., is proportional to $\vec{b} \times \vec{c}$. Taking the proportionality constant to be α,

$$\vec{a}^* = \alpha (\vec{b} \times \vec{c}),$$

and determining the value of α by $\vec{a}^* \cdot \vec{a} = \vec{a} \cdot \vec{a}^* = 1$ yields

$$\alpha (\vec{a} \cdot \vec{b} \times \vec{c}) = \alpha V_{cell} = 1,$$

or $\alpha = V_{cell}^{-1}$. Proceeding similarly for $\vec{b}^*$ and $\vec{c}^*$ yields

$$\vec{a}^* = \frac{\vec{b} \times \vec{c}}{V_{cell}}, \quad \vec{b}^* = \frac{\vec{c} \times \vec{a}}{V_{cell}}, \quad \vec{c}^* = \frac{\vec{a} \times \vec{b}}{V_{cell}},$$

and with $\vec{K} = m^* \vec{a}^* + n^* \vec{b}^* + p^* \vec{c}^*$ the functions $\sin 2\pi\vec{K} \cdot \vec{r}$, $\cos 2\pi\vec{K} \cdot \vec{r}$ and $\exp 2\pi i \vec{K} \cdot \vec{r}$ all have the periodicity of the lattice (each is a basis function for the totally symmetric irr. rep.). The definition of $\vec{K}$ is such that $\vec{K} \cdot \vec{r}$ is dimensionless, thus $\vec{K}$ has the dimension of reciprocal length and is called a reciprocal vector or, since m^*, n^* and p^* can be allowed all integral values thereby generating a lattice (a reciprocal lattice), $\vec{K}$ is called a reciprocal lattice vector.

The essential feature used in the definition of the reciprocal lattice vector is that $\vec{K} \cdot \vec{t}$ = integer (note: it is also common to define $\vec{K}$ such that $\vec{K} \cdot \vec{t}$ is an integral multiple of 2π, and this reciprocal vector is obtained from that defined here by multiplication by 2π). It follows that if $\beta | \vec{t}$ is a symmetry operation of a crystalline solid then $\vec{K} \cdot \beta\vec{t}$ is an integer if $\vec{K} \cdot \vec{t}$ is. However if $\vec{K} \cdot \beta\vec{t}$ is an integer then so too is $\beta^{-1}\vec{K} \cdot \vec{t}$. This is true for a given $\vec{t}$ and all of the reciprocal vectors generated by the rotational part of the symmetry operations of the space group. Thus if $\vec{K}$ is a reciprocal lattice vector so too is $\beta\vec{K}$ and the reciprocal lattice has the same rotational symmetry as the real lattice.

III. RECIPROCAL SPACE

It becomes necessary when one wishes to consider periodicities different from that of the lattice, i.e., to consider basis functions for irr. reps. of the translational subgroup other than the totally symmetric, to generalize the reciprocal vector concept to include vectors with nonintegral components of $\vec{a}^*$, $\vec{b}^*$ and/or $\vec{c}^*$. Therefore the <u>wave vector</u>, $\vec{k}$, which is a general, nonintegral component vector in <u>reciprocal space</u> is introduced. For example, $\mu\vec{a}^*$ corresponds to a wave vector along the $\vec{a}^*$ direction.

We have shown that among the rotational symmetry operations of the reciprocal lattice are included those which make up the crystal class of the space group. These rotational symmetry operations carry wave vectors into symmetrically equivalent wave vectors and thus are symmetry operations of reciprocal space. The general wave vectors, under the β operations of the crystal class, fall into three cases. Either: 1. they transform into different wave vectors which do not differ from the original by a reciprocal lattice vector, or 2. they transform into different wave vectors which do differ from the original by a reciprocal lattice vector, or 3. they transform into themselves. For example, $\mu\vec{a}^*$ with $0 < \mu < 1/2$ transforms into itself under C_{2x} or σ_z (case 3) but into $-\mu\vec{a}^*$ under i and C_{2z} (case 1). On the other hand, $1/2\,\vec{a}^*$ transforms into $-1/2\,\vec{a}^*$ under i and C_{2z} and thus $\vec{a}^*/2$ falls under case 2 for these operations.

Cases 2 and 3 can be included in the same category if it is specified that symmetry operations in this category carry $\vec{k}$ into itself modulo a reciprocal lattice vector, i.e., $\vec{k}$ into $\vec{k} + \vec{K}$ including $\vec{K} = 0$. The set of all β parts of the symmetry operations that carry $\vec{k}$ into $\vec{k}$ modulo $\vec{K}$ form a point group called the <u>point group of the wave vector</u>, $g_0(\vec{k})$. This statement is proven below. The remaining β parts of the space group operations carry the $\vec{k}$ vector into other $\vec{k}$ vectors; those which do not differ by a reciprocal lattice vector are said to form a star.

For example, consider the wave vector $\vec{a}^*/2$ in the case of a hexagonal space group in the class D_{6h} (Fig. 15). The vectors that result from the β operations on $\vec{a}^*/2$ are $\pm\vec{a}^*/2$, $\pm\vec{b}^*/2$ and $\pm(\vec{a}^* - \vec{b}^*)/2$. Of these $\vec{a}^*/2$, $\vec{b}^*/2$ and $(\vec{a}^*-\vec{b}^*)/2$ form a star. Those symmetry operations that carry $\vec{a}^*/2$ into $\pm\vec{a}^*/2$ (into $\vec{a}^*/2$ modulo $\vec{a}^*$) are ε, C_{2z}, C_{2y}, $C_{2(2x+y)}$, i, σ_z, σ_y and σ_{2x+y}, i.e., the symmetry operations of D_{2h} expressed in hexagonal coordinates. Thus the point group of the wave vector $\vec{a}^*/2$ is D_{2h}. The point groups of the other wave vectors in the star $\vec{b}^*/2$ and $(\vec{a}^* - \vec{b}^*)/2$, are also D_{2h}, but these point groups consist of different sets of two-fold rotations and reflections than those appropriate to $\vec{a}^*/2$ (and, of course, these two sets also differ from each other).

To prove that the set of β operations of the crystal class that carry $\vec{k}$ into $\vec{k}$ modulo $\vec{K}$ form a group in general we must prove closure of the set, that the set contains all inverses and an identity and that the binary operation is associative. First, regarding closure, let

$$\beta_1\vec{k} = \vec{k} + \vec{K}_1$$

and

$$\beta_2\vec{k} = \vec{k} + \vec{K}_2 .$$

Then

$$\beta_1\beta_2\vec{k} = \beta_1\vec{k} + \beta_1\vec{K}_2$$
$$= \vec{k} + \vec{K}_1 + \beta_1\vec{K}_2$$

and from the fact that $\beta_1 \vec{K}_2$ is necessarily a reciprocal lattice vector, and

thus $\vec{K}_1 + \beta \vec{K}_2$ is, we can write

$$\beta_1 \beta_2 \vec{k} = \vec{k} + \vec{K}_3$$

where $\vec{K}_3 = \vec{K}_1 + \beta \vec{K}_2$ and closure is proven. Next consider the existence of an

identity;

$$\epsilon \vec{k} = \vec{k} + \vec{K}$$

with $\vec{K} = 0$ and thus ϵ is in the set. Regarding inverses, if

$$\beta \vec{k} = \vec{k} + \vec{K}$$

then

$$\beta^{-1} \beta \vec{k} = \vec{k} = \beta^{-1} \vec{k} + \beta^{-1} \vec{K}$$

or

$$\beta^{-1} \vec{k} = \vec{k} - \beta^{-1} \vec{K}$$

and since $-\beta^{-1} \vec{K}$ is necessarily a reciprocal lattice vector we can set $\vec{K}' = -\beta^{-1} \vec{K}$, and

$$\beta^{-1} \vec{k} = \vec{k} + \vec{K}'$$

showing that all β's have inverses within the set. Associativity is a
property of the binary combination operation in the crystal class and thus
this operation is associative. Hence the set of all operations of the
crystal class that carry $\vec{k}$ into $\vec{k}$ modulo $\vec{K}$ form a group $g_0(\vec{k})$.

The reason for introducing the concept of reciprocal space is that
it provides access to basis functions and representations of the
translational symmetry operations. So far as the translational subgroup is
concerned, exp $2\pi i \vec{k} \cdot \vec{r}$ is a basis function for the translational symmetry
operations, i.e., transforming $\vec{r}$ into $\vec{r} + \vec{T}$ transforms exp $2\pi i \vec{k} \cdot \vec{r}$ into exp
$2\pi i \vec{k} \cdot (\vec{r} + \vec{T})$ = exp $2\pi i \vec{k} \cdot \vec{T}$ exp $2\pi i \vec{k} \cdot \vec{r}$ and thus into a constant (exp $2\pi i \vec{k} \cdot \vec{T}$)
times itself.

For example if $\vec{k} = \vec{a}^*/2$ then exp $2\pi i \vec{k} \cdot \vec{r}$ = exp $\pi i x$ and this function is transformed into exp $\pi i(x+m)$ = exp $m\pi i$ exp $\pi i x$ by the pure translation $\vec{T} = m\vec{a} + n\vec{b} + p\vec{c}$, and is thus symmetric with respect to translations with m even (exp $m\pi i$ = 1) and antisymmetric with respect to translations with m odd (exp $m\pi i$ = -1).

If we consider, on the other hand, transforming $\vec{k}$ into $\vec{k}+\vec{K}$ then exp $2\pi i \vec{k} \cdot \vec{r}$ transforms into exp $2\pi i(\vec{k}+\vec{K}) \cdot \vec{r}$ = exp $2\pi i \vec{k} \cdot \vec{r}$ exp $2\pi i \vec{K} \cdot \vec{r}$, i.e., into the product of the function with a totally symmetric function. So far as characterizing functions by symmetry is concerned, these functions are basis functions for the same irr. rep. That is, for example, both exp $(2\pi i \vec{a}^*/2) \cdot \vec{r}$ = exp $\pi i x$ and exp $2\pi i$ $(\vec{a}^*/2 + \vec{K}) \cdot \vec{r}$ = exp $2\pi i \vec{K} \cdot \vec{r}$ exp $\pi i x$ are basis functions for the irr. rep. of the translational subgroup that is symmetric with respect to even $\vec{a}$ and antisymmetric with respect to odd $\vec{a}$ translations. The β operations that carry $\vec{k}$ into $\vec{k}$ modulo $\vec{K}$ carry a basis function for an irr. rep. of the translational subgroup into a different basis function for the same irr. rep. of the translations and therefore these β operations are considered together in the group $g_o(\vec{k})$.

What has been said above means that functions of all possible translational symmetries are considered when one considers the set of all wave vectors defined so that $\vec{k}$ modulo $\vec{K}$ does not belong to the set. This fact leads to the definition of the Brillouin zone. The first Brillouin zone is the volume of reciprocal space containing all wave vectors from the origin (called point Γ) such that $\vec{k}$ modulo $\vec{K}$ does not belong to the set plus (in order to get a closed set) all vectors terminating on the boundary of the set (even though some of the vectors terminating on the boundary may differ by $\vec{K}$). In order to construct this zone it is only necessary to enclose a volume about a reciprocal lattice point (taken as point Γ) by constructing planes which are perpendicular bisectors of all lines connecting lattice points to the origin as shown for a monoclinic reciprocal lattice in projection in Fig. 16. Note that $\pm\, \vec{a}^*/2$, $\pm\, \vec{b}^*/2$ and $\pm\, (\vec{a}^*-\vec{b}^*)/2$ all terminate on the zone boundary and are therefore included within the zone even though the members of each pair differ by a reciprocal lattice vector.

This fact gives rise to special symmetries for some points on the boundary of the zone. For example, in the monoclinic case (Fig. 16), assuming the crystal class to be C_{2h}, μa^* with $0 < \mu < 1/2$ is carried into itself under σ_z, but not i and C_{2z}, and thus $\{\varepsilon,\ \sigma_z\}$ constitutes $g_0(\mu\vec{a}^*)$. On the other hand, when $\mu = 1/2$ both i and C_{2z} carry $\vec{a}^*/2$ into $-\vec{a}^*/2$, which differs from $\vec{a}^*/2$ by $\vec{a}^*$, and thus both C_{2z} and i are in the point group of the wave vector $\vec{a}^*/2$, i.e., $g_0(\vec{a}^*/2)$ is C_{2h}.

IV. IRREDUCIBLE REPRESENTATION OF SPACE GROUPS

The irr. reps. of space groups are, for a large number of cases, simply related to the irr. reps. of the point groups $g_0(\vec{k})$. In order to discover these cases and find the irr. reps. consider as a possible irr. rep. for $\beta_i|\vec{t}_i$ at point $\vec{k}$ in the first Brillouin zone

$$\Gamma(\beta_i|\ \vec{t}_i) = e^{2\pi i \vec{k}\cdot\vec{t}_i}\ \Gamma(\beta_i)$$

where $\Gamma(\beta_i)$ is the irr. rep. of β_i in $g_0(\vec{k})$, and as has just been shown, exp $2\pi i \vec{k}\cdot\vec{t}_i$, is the irr. rep. of $\vec{t}_i$ when $\vec{t}_i$ is in the group of the pure translations (which is not always the case). It is because of this last parenthetical remark that the trial irr. rep. might not work. If such a form is to be a representation then

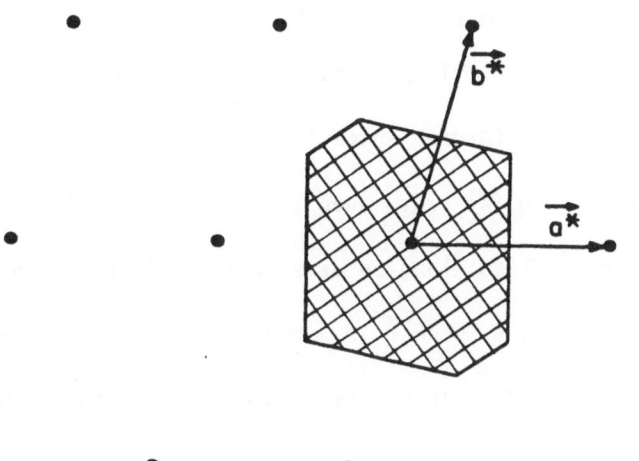

Fig. 16. A monoclinic reciprocal lattice projected along the unique axis showing the projected first Brillouin zone. The zone extends to $\pm \vec{c}^{*}/2$ relative to the $\vec{a}^{*} - \vec{b}^{*}$ plane.

$$\Gamma(\beta_2 \mid \vec{t}_2)\ \Gamma\ (\beta_1 \mid \vec{t}_1) = \Gamma\ (\beta_1 \beta_2 \mid \beta_2 \vec{t}_1 + \vec{t}_2)$$

i.e.,

$$e^{2\pi i \vec{k} \cdot \vec{t}_2}\ \Gamma\ (\beta_2)\ e^{2\pi i \vec{k} \cdot \vec{t}_1}\ \Gamma\ (\beta_1) = e^{2\pi i \vec{k} \cdot (\beta_2 \vec{t}_1 + \vec{t}_2)}\ \Gamma\ (\beta_2 \beta_1)\ .$$

Since $\Gamma(\beta_2)\Gamma(\beta_1) = \Gamma(\beta_2\beta_1)$, and since exp $2\pi i \vec{k} \cdot \vec{t}_2$ can be cancelled from both sides, the equality holds if and only if

$$\vec{k} \cdot \vec{t}_1 = \vec{k} \cdot \beta \vec{t}_1$$

for all space group operations with rotational parts in $g_0(\vec{k})$. This equality can be seen to be valid for a number of cases: 1. $\vec{k} = 0$, i.e., at point Γ for all space groups, 2. $\vec{t}_i = \vec{t}_i$ for all operations with rotational parts in $g_0(\vec{k})$ e.g., at all points of reciprocal space for symmorphic space groups, 3. when $\beta\vec{t}_1 = \vec{t}_1$ for all β in $g_0(\vec{k})$ e.g., when $g_0(\vec{k}) = \epsilon$ (general points of all space groups) and when $\vec{t}_i$ lies along an axis or in a plane and there are no intersecting axes or planes, 4. when $\vec{k}$ is perpendicular to $\vec{t}_i$ and $\beta\vec{t}_i$ e.g., when $\vec{k} = \vec{a}^*/2$ and $\vec{t} = \vec{c}/2$. These cases cover most of the points in most Brillouin zones. The cases that are not covered are those, for example, for which $\vec{k}$ and $\vec{t}$ are parallel (e.g., $\vec{k} = \vec{c}^*/2$ and $\vec{t} = \vec{c}/2$) for a nonsymmorphic space group such as $P2_1/m$. In cases such as these the irr. reps. suggested above need not work and, in fact, as we shall show, do not work. Therefore a different approach to obtaining irr. reps. at these points is needed.

An approach to this problem, motivated by the above, is to define what is called a "loaded" representation (which is, in fact, not necessarily a representation at all) for β_i in $g_0(\vec{k})$ according to

$$\tilde{\Gamma}(\beta_i) = e^{-2\pi i \vec{k} \cdot \vec{t}_i} \ \Gamma(\beta_i \mid \vec{t}_i)$$

and we note that $\tilde{\Gamma}(\beta_i)$ is in fact $\Gamma(\beta_i)$ in the cases 1-4 listed above. A multiplication table for the $\tilde{\Gamma}$'s can be generated, for since $\Gamma(\beta_i \mid \vec{t}_i)$ must multiply like the symmetry operations,

$$\tilde{\Gamma}(\beta_2)\,\tilde{\Gamma}(\beta_1) = \epsilon^{-2\pi i \vec{k}\cdot(\vec{t}_2+\vec{t}_1)}\ \Gamma(\beta_2\,|\,\vec{t}_2)\ \Gamma(\beta_1\,|\,\vec{t}_1)$$

$$= \epsilon^{-2\pi i \vec{k}\cdot(\vec{t}_2+\vec{t}_1)}\ \Gamma(\beta_2\beta_1\,|\beta_2\vec{t}_1+\vec{t}_2)$$

$$= \epsilon^{2\pi i \vec{k}\cdot(\beta_2\vec{t}_1-\vec{t}_1)}\ \tilde{\Gamma}(\beta_2\beta_1).$$

For example in the case of $P2_1/m$ at $\vec{c}^*/2$ the essential symmetry operations are $\epsilon|0$, $C_{2z}\,|\,\vec{c}/2$, $i\,|\,0$, $\sigma_z\,|\,\vec{c}/2$. Adopting the convention that the symmetry operations listed along the top of a table correspond to those operations performed first ($\beta_1\,|\,\vec{t}_1$ in the above) and those listed down the side to those performed second ($\beta_2|\vec{t}_2$), the multiplication table (Table 5) for the loaded reps. can be constructed. Realizing that $\tilde{\Gamma}(\beta_2)\,\tilde{\Gamma}(\beta_1)$ = $\tilde{\Gamma}(\beta_2\beta_1)$ when $t_1 = 0$ allows all of the entries in the ϵ and i columns to be directly introduced as $\tilde{\Gamma}(\beta_2\beta_1)$, the same is true when β_2 carries t_1 into itself, i.e., for the rows led by ϵ and C_{2z}. This leaves four entries, $\tilde{\Gamma}(i)\,\tilde{\Gamma}(\sigma_z)$, $\tilde{\Gamma}(i)\,\tilde{\Gamma}(C_{2z})$, $\tilde{\Gamma}(\sigma_z)\,\tilde{\Gamma}(C_{2z})$ and $\tilde{\Gamma}(\sigma_z)\,\tilde{\Gamma}(\sigma_z)$ to be determined . In all of these cases $\beta_2\vec{t}_1 = -\vec{t}_1$ ($=-\vec{c}/2$) and thus

$$\exp 2\pi i\ \frac{\vec{c}^*}{2}\cdot(-\frac{\vec{c}}{2}-\frac{\vec{c}}{2}\ ) = e^{-\pi i} = -1,$$

and thus $\tilde{\Gamma}(\beta_2)\,\tilde{\Gamma}(\beta_1) = -\tilde{\Gamma}(\beta_2\beta_1)$ in these cases. The completed table is as shown (Table 5) and it presents two noteworthy features. First of these is that not all of the loaded reps. commute, e.g.,

$$\tilde{\Gamma}(i)\,\tilde{\Gamma}(\sigma_z) = -\tilde{\Gamma}(\sigma_z)\,\tilde{\Gamma}(i)$$

TABLE 5. THE MULTIPLICATION TABLE FOR A LOADED REP. OF $P2_1/m$ at $\vec{c}^{*}/2$.

β_2 \ β_1	$\tilde{\Gamma}(\varepsilon)$	$\tilde{\Gamma}(C_{2z})$	$\tilde{\Gamma}(i)$	$\tilde{\Gamma}(\sigma_z)$
$\tilde{\Gamma}(\varepsilon)$	$\tilde{\Gamma}(\varepsilon)$	$\tilde{\Gamma}(C_{2z})$	$\tilde{\Gamma}(i)$	$\tilde{\Gamma}(\sigma_z)$
$\tilde{\Gamma}(C_{2z})$	$\tilde{\Gamma}(C_{2z})$	$\tilde{\Gamma}(\varepsilon)$	$\tilde{\Gamma}(\sigma_z)$	$\tilde{\Gamma}(i)$
$\tilde{\Gamma}(i)$	$\tilde{\Gamma}(i)$	$-\tilde{\Gamma}(\sigma_z)$	$\tilde{\Gamma}(\varepsilon)$	$-\tilde{\Gamma}(C_{2z})$
$\tilde{\Gamma}(\sigma_z)$	$\tilde{\Gamma}(\sigma_z)$	$-\tilde{\Gamma}(i)$	$\tilde{\Gamma}(C_{2z})$	$-\tilde{\Gamma}(\varepsilon)$

and thus the $\tilde{\Gamma}$'s cannot be numbers (real or complex), but rather must be at least 2x2 matrices. The second feature of note is that no inverse for $\tilde{\Gamma}(\sigma_z)$ is present in the table, i.e., the $\tilde{\Gamma}$'s are not a representation of a group. Nonetheless the $\tilde{\Gamma}$'s are useful, as shown in what follows.

Since we know that there is no one-dimensional solution to the multiplication table we seek the solution in 2x2 matrices. We set $\tilde{\Gamma}(\varepsilon) = \begin{pmatrix} 1 & 0 \\ 0 & 1 \end{pmatrix}$ and $\tilde{\Gamma}(\sigma_z) = \begin{pmatrix} a & b \\ c & d \end{pmatrix}$ and from $\tilde{\Gamma}(\sigma_z)\,\tilde{\Gamma}(\sigma_z) = -\tilde{\Gamma}(\varepsilon)$ we find that

$$a^2 + bc = d^2 + bc = -1$$

and

$$b(a+d) = c(a+d) = 0.$$

A solution is $b = -c = 1$ and $a = d = 0$, Letting $\tilde{\Gamma}(i) = \begin{pmatrix} e & f \\ g & h \end{pmatrix}$ and using

$$\begin{pmatrix} 0 & 1 \\ -1 & 0 \end{pmatrix} \begin{pmatrix} e & f \\ g & h \end{pmatrix} = - \begin{pmatrix} e & f \\ g & h \end{pmatrix} \begin{pmatrix} 0 & 1 \\ -1 & 0 \end{pmatrix},$$

where $\begin{pmatrix} 0 & 1 \\ -1 & 0 \end{pmatrix}$ is the matrix found for $\tilde{\Gamma}(\sigma_z)$, and

$$\begin{pmatrix} ef \\ gh \end{pmatrix} \begin{pmatrix} ef \\ gh \end{pmatrix} = \begin{pmatrix} 10 \\ 01 \end{pmatrix}$$

yields, from the former g=f and -h=e and from the latter

$$e^2 + fg = h^2 + fg = 1$$

and

$$f(e+h) = g(e+h) = 0,$$

and thus a solution is f=g=1 or $\tilde{\Gamma}(i) = \begin{pmatrix} 01 \\ 10 \end{pmatrix}$. It then follows that

$$\tilde{\Gamma}(C_{2z}) = \begin{pmatrix} 10 \\ 0\text{-}1 \end{pmatrix},$$

and thus we have

β	ε	C_{2z}	i	σ_z
$\tilde{\Gamma}(\beta)$	$\begin{pmatrix} 10 \\ 01 \end{pmatrix}$	$\begin{pmatrix} 10 \\ 0\text{-}1 \end{pmatrix}$	$\begin{pmatrix} 01 \\ 10 \end{pmatrix}$	$\begin{pmatrix} 01 \\ \text{-}10 \end{pmatrix}$

.

Using $\Gamma(\beta \mid \vec{t}) = \exp 2\pi i \vec{k} \cdot \vec{t} \, \tilde{\Gamma}(\beta)$ we obtain the small representation of $P2_1/m$ at $\vec{c}^*/2$:

$\beta \mid \vec{t}$	$\varepsilon \mid 000$	$C_{2z} \mid 00^{1}/_{2}$	$i \mid 000$	$\sigma_z \mid 00^{1}/_{2}$
$\Gamma(\beta \mid \vec{t})$	$\begin{pmatrix} 10 \\ 01 \end{pmatrix}$	$\begin{pmatrix} i0 \\ 0\text{-}i \end{pmatrix}$	$\begin{pmatrix} 01 \\ 10 \end{pmatrix}$	$\begin{pmatrix} 0i \\ \text{-}i0 \end{pmatrix}$

from which the irr. rep. of the space group can be generated.

Note that the functions $\exp \pi i z = \phi_1$ and $\exp -\pi i z = \phi_2$ are basis functions for this irr. rep. (Table 6).

Thus we are able to straightforwardly obtain space group irr. reps. from those of $g_0(\vec{k})$ in a great many cases, and to find loaded irr. reps. from which those of the space group can be inferred in others. The irr. reps. have the physical significance that properties or changes that correspond to a function of position in the crystal that is a basis function, or a linear combination of basis functions, for a given irr. rep. are related by symmetry to other properties or changes which correspond to other combinations of those basis functions, but are unrelated to those properties or changes which correspond to functions which are combinations of basis functions for other irr. reps. This fact has been exploited by Landau in the theory concerning symmetry aspects of second-order phase transitions that bears his name and is a major topic of the remainder of this book.

TABLE 6.

$\epsilon \mid 000$	x, y, z	$\begin{pmatrix}1 & 0 \\ 0 & 1\end{pmatrix}\begin{pmatrix}e^{\pi i z} \\ e^{-\pi i z}\end{pmatrix} = \begin{pmatrix}e^{\pi i z} \\ e^{-\pi i z}\end{pmatrix}$
$C_{2z} \mid 00\,{}^1/_2$	$\bar{x}, \bar{y}, z + {}^1/_2$	$\begin{pmatrix}i & 0 \\ 0 & -i\end{pmatrix}\begin{pmatrix}e^{\pi i z} \\ e^{-\pi i z}\end{pmatrix} = \begin{pmatrix}e^{\pi i(z+{}^1/_2)} \\ e^{-\pi i(z+{}^1/_2)}\end{pmatrix}$
$i \mid 000$	$\bar{x}, \bar{y}, \bar{z}$	$\begin{pmatrix}0 & 1 \\ 1 & 0\end{pmatrix}\begin{pmatrix}e^{\pi i z} \\ e^{-\pi i z}\end{pmatrix} = \begin{pmatrix}e^{-\pi i z} \\ e^{\pi i z}\end{pmatrix}$
$\sigma_z \mid 00\,{}^1/_2$	$x, y, \bar{z} + {}^1/_2$	$\begin{pmatrix}0 & i \\ -i & 0\end{pmatrix}\begin{pmatrix}e^{\pi i z} \\ e^{-\pi i z}\end{pmatrix} = \begin{pmatrix}e^{\pi i(\bar{z}+{}^1/_2)} \\ e^{-\pi i(\bar{z}+{}^1/_2)}\end{pmatrix}$

SECOND-ORDER PHASE TRANSITIONS

I. INTRODUCTION

This chapter treats the thermodynamic and group theoretical techniques useful in the consideration of phase transitions which occur without the coexistence of two phases, i.e., without the nucleation and growth of a new phase. The transitions under consideration occur with a change of symmetry at a certain thermodynamic state during a continuous structure change through that state. Such structure changes may be of three types: order-disorder, displacive and a combination of order-disorder and displacive. In the latter case the ordering is incommensurate with the translational periodicity of the lattice, as will be shown in a later section.

The classical example of a pure order-disorder transition that occurs with a continuous change in structure is that in CuZn which is called the β-β' brass transition (1). At very low temperatures there is a tendency for CuZn to adopt the CsCl structure (Fig. 17), i.e., for the Cu atoms to preferentially occupy a simple cubic lattice with origin at 0,0,0 and for the Zn atoms to occupy a similar lattice with origin at $1/2$, $1/2$, $1/2$ (or, what is equivalent by a change of origin of the structure, vice versa). As the temperature is increased there is a tendency for the atoms to interchange at random, an interchange which increases the configuration entropy of the system ($\Delta S = R\ln 2$ for the change from CsCl-type to completely random distribution of Cu and Zn atoms on a bcc lattice).

An order-disorder parameter, η, can be defined

$$\eta = 2f_{Cu} - 1$$

where f_{Cu} is the fraction of 0,0,0 sites occupied by Cu atoms. This parameter is 0 for the random distribution and 1 for the completely ordered CsCl-type structure. From the point of view of symmetry there are symmetry elements present in the $\eta = 0$ case that are missing when $\eta \neq 0$, for example translation by $^1/_2$, $^1/_2$, $^1/_2$ (i.e., $(\vec{a} + \vec{b} + \vec{c})/2$) is a symmetry operation only when $\eta = 0$. It is experimentally observed that η does vanish continuously at sufficiently high temperatures and is, as stated above, nonzero at low temperatures. Since a symmetry operation can be lost (or gained) only at a state point it follows that a phase transition can be defined as a symmetry change, and a transition point (e.g., temperature) as the state which separates $\eta = 0$ from $\eta \neq 0$ structures (provided the order-disorder process takes place continuously, as apparently is the case).

Some other examples of apparently continuous order-disorder processes are the ordering of vacancies in the Sc substructure of $Sc_{1-x}S$ in alternate metal planes along the (1,1,1) direction of the defect NaCl-type structure (a change from Fm3m to R$\bar{3}$m symmetry (2)) and the ordering of Cr vacancies in defect $Cr_{1-x}S$ in alternate planes along the c-axis of the defect hexagonal (P6$_3$/mmc)NiAs-type to yield the defect CdI_2-type (P$\bar{6}$m2) structure, (3). A further ordering of the vacancies within the a-b plane is also thought to occur.

An example of a displacive transition which is believed to occur continuously is the NiAs-type to MnP-type phase transition in CoAs (4) and VS (5). In this case the atoms are observed to move in a concerted fashion such as to destroy the hexagonal symmetry (P6$_3$/mmc) but maintain orthorhombic symmetry (Pcmn). The movement of the metal atoms in the hexagonal layer at z = 0 is shown schematically in Fig. 18. In this case η can be defined as

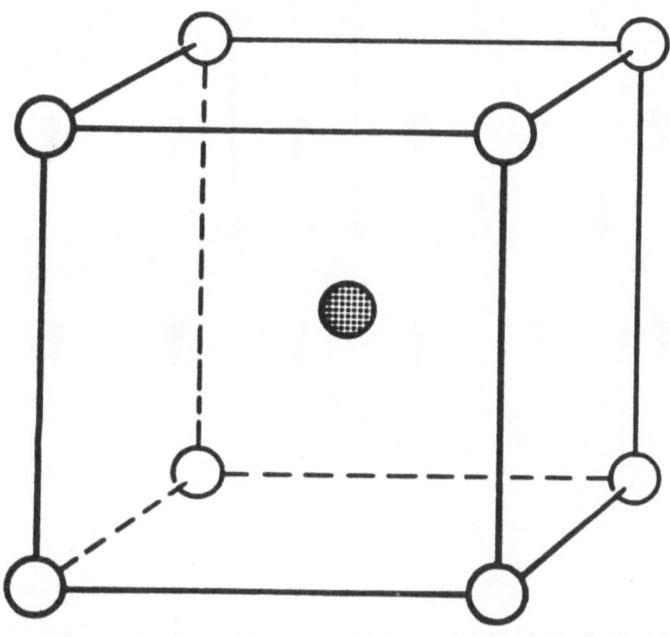

Fig. 17. The CsCℓ-type structure. Atom of type A at 0,0,0 and of type B at 1/2,1/2,1/2. Random occupation yields bcc.

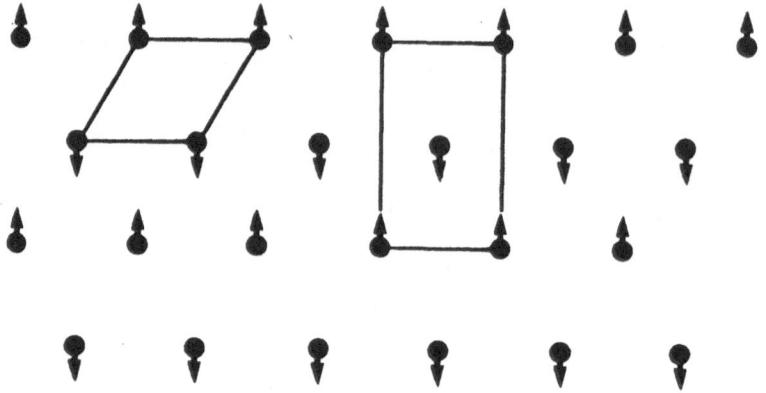

Fig. 18. Metal atom positions in the z = o metal atom layer of the NiAs-type structure with the distortion to MnP-type observed for VS indicated by the arrows. The metal atoms at $z = 1/2$ move in the opposite direction. The hexagonal (NiAs) and orthorhombic (MnP) cells are outlined.

the length of the vectors describing the distortion, and with normalization this η varies between 1 at 0 K and 0 at the transition temperature.

The third type of continuous phase transition combines displacive and order-disorder changes and therefore leads necessarily to an incommensurate ordering as has recently been described (6). This case will be discussed later, after the Landau theory has been developed.

II. THERMODYNAMICS OF SECOND-ORDER PHASE TRANSITIONS

At first, in order to appreciate the nature of the phase transitions under discussion, it is desirable to understand the thermodynamic consequences of the proposed phenomena. The phenomena have in common the following description: in some region of thermodynamic space the sample has one set of symmetry elements and in another region a different set, however efforts to discover states at which samples with the two sets coexist in equilibrium fail and it is concluded that the transition occurs without this coexistence. A phase diagram in this case, with one symmetry labeled α and the other β, and with the variables T and X (mole fraction) is shown in Fig. 19a.

We introduce the Gibbs-Konovalow (G-K) equation (7)

$$\left(\frac{\partial T}{\partial X_A^\alpha} \right)_P = - \frac{\{ \left(\frac{\partial \mu_A^\alpha}{\partial X_A^\alpha} \right)_{T,P} + \left(\frac{\partial \mu_B^\alpha}{\partial X_B^\alpha} \right)_{T,P} \} \Delta X}{\Delta \bar{S} + (\bar{S}_A^\alpha - \bar{S}_B^\alpha) \, \Delta X}$$

where $\left(\frac{\partial T}{\partial X_A^\alpha} \right)_P$ is the slope of a phase boundary on a T-X diagram (see

Fig. 19b), $\left(\frac{\partial \mu_A^\alpha}{\partial X_A^\alpha} \right)_{T,P}$ is the mole fraction partial derivative of the

chemical potential in a phase (α) bounded by the phase boundary, ΔX is the difference in the mole fraction of A in the two phases separated by the two-phase region enclosed by the boundary and $\Delta \bar{S}$ and $\bar{S}_A^\alpha$ are the difference in molar entropy of the phases and the partial molar entropy of A in α,

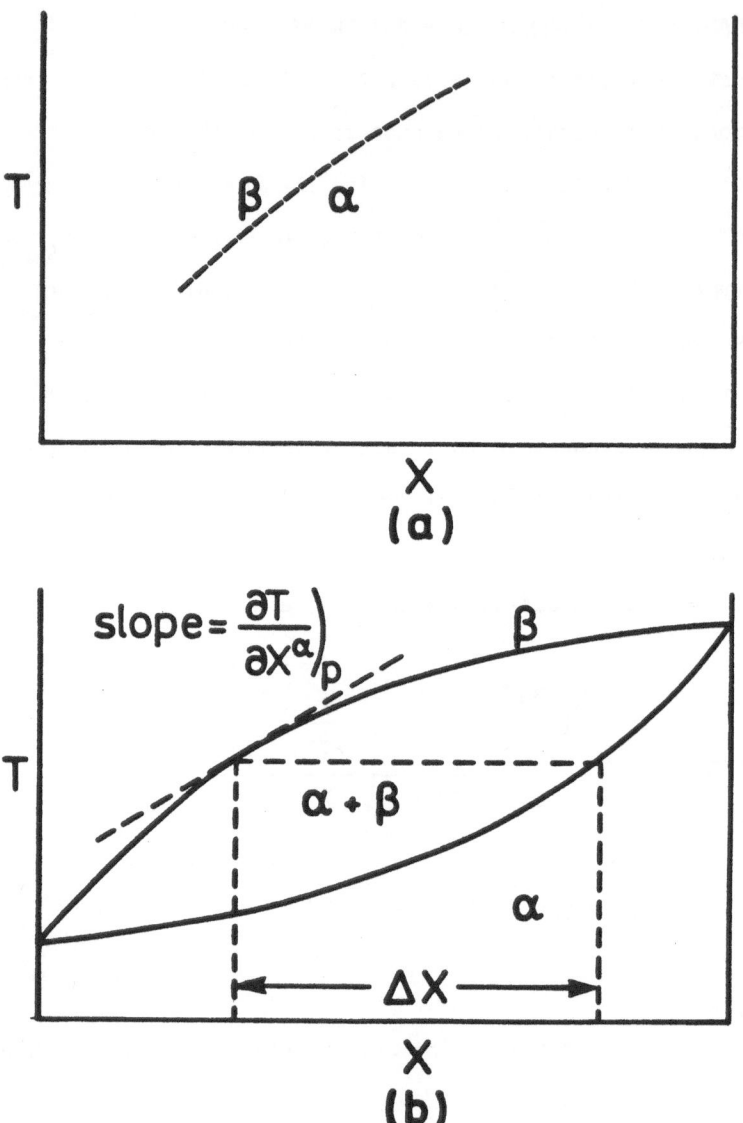

Fig. 19. Schematic T-x phase diagrams. (a) continuous structure change (b) two phases coexist in equilibrium.

respectively. We remark that this equation is the differential equation of
the lines that appear on binary phase diagrams and it provides the general
theoretical description of such lines.

Applying this equation to the case in question we must first note
that $\left(\frac{\partial\mu}{\partial X}\right)_{T,P}$ is not infinite (for ideal solutions it is RT/X) except at the
X=0 boundaries of phase diagrams. Thus for changes in symmetry occurring
with continuous changes in structure in nonstoichiometric compounds, the fact
that $\Delta X = 0$ and $\left(\frac{\partial T}{\partial X_A^\alpha}\right)_P \neq 0$ (see Fig. 19) requires that $\Delta S = 0$, i.e., that
the entropy changes continuously at the transition point. Note also that the
continuity of the process means also that $\Delta V = 0$ (the volume changes
continuously). Processes for which ΔS and ΔV are zero are called second-
order phase transitions.

It is helpful in considering first-order phase transitions to
consider a G-T diagram (at constant pressure and composition) such as shown
in Fig. 20. In this case α corresponds to a phase stable at $T < T_t$
(since $\Delta G^{\alpha \to \beta} > 0$ here) and β to a phase stable at $T > T_t$ (since $\Delta G^{\alpha \to \beta} < 0$
here), and α and β coexist in equilibrium at T_t (where $\Delta G = o$). Because
$\left(\frac{\partial G}{\partial T}\right)_P = -S$ it follows that $\Delta S \neq 0$ because the curves intersect and cross
i.e., they have different slopes.

If an attempt is made to apply the same thinking to the second-order
case some apparent difficulties arise. For example suppose we expand G in T
about T_t:

$$G = G^\circ + \frac{\partial G}{\partial T}\Big]_{T=T_t} (T-T_t) + 1/2 \frac{\partial^2 G}{\partial T^2}\Big]_{T=T_t} (T-T_t)^2 + \ldots$$

or

$$G = G^\circ - S^\circ (T-T_t) - 1/2 \frac{C_p^\circ}{T_t} (T-T_t)^2 + \ldots$$

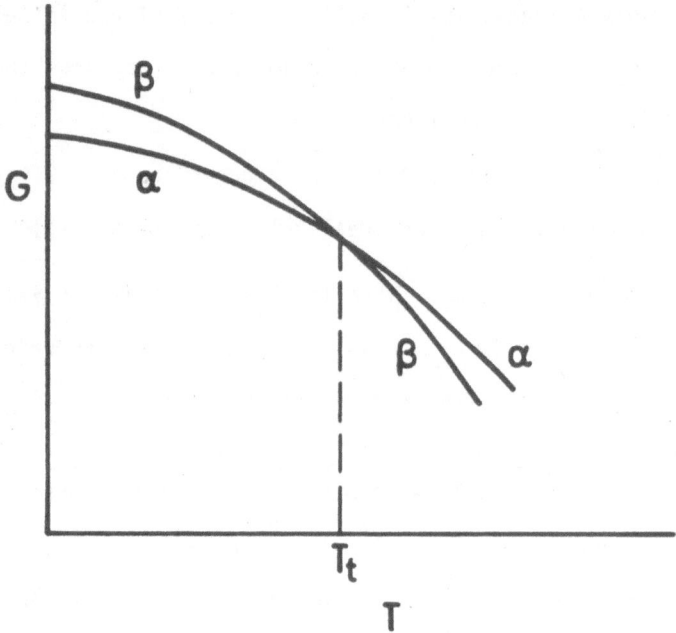

Fig. 20. Schematic G vs. T (at constant P,X) phase diagram. A first-order phase transition occurs at $T=T_t$.

where the super zero means "evaluated at T_t". Applying this equation to α and β, taking the difference, and recognizing that ΔG° and ΔS° both vanish for a second-order phase transition, yields

$$\Delta G \cong - \frac{\Delta C_p^\circ}{2T_t} (T-T_t)^2$$

for sufficiently small values of $T-T_t$. But this equation implies that ΔG is an even function of $T-T_t$ in the neighborhood of T_t and therefore the same form must be stable for both $T > T_t$ and $T < T_t$ (see Fig. 21). This contradicts the notion of a phase transition.

The contradiction is eliminated when it is recognized that the α symmetry cannot exist for $T > T_t$ ($T < T_t$ is discussed in a later section). As the temperature increases from $T = 0$ to $T = T_t$ η approaches zero and vanishes at T_t. The physical meaning of a continuation of η to negative values is an ordering or distortion which inverts that for $\eta > 0$. For example, the continuation of η to negative values in the CuZn case would have the meaning of ordering Zn atoms at the original Cu positions and in the NiAs-type to MnP-type case would have the meaning of inverting the arrows of Fig. 18. In either case continuation to $\eta < 0$ would correspond to a return to the structure stable at $T < T_t$, but with a change of origin. It follows that the expansion of G in T on both sides of T_t was not meaningful for symmetry α -- only the side of T_t corresponding to $\eta > 0$ is physically meaningful.

There remains another distinction that can be made between the G vs. T diagrams appropriate to first- and second-order phase transitions. This has to do with the area between the α and β curves which, in the second-order case, is a region that is in principle accessible to the system. The β curve for $T < T_t$ corresponds to metastable $\eta = 0$, a possible physical state in the order-disorder case because there is an energy barrier to diffusion, but only theoretically accessible in the displacive case because there is no barrier to distortion. However the choice of the $\eta = 0$ curve for continuation to $T < T_t$ is arbitrary -- any $\eta \neq 0$ curve could also be continued. Said in a different way, because of the continuity of the process there is a continuity of states accessible to the system corresponding to the area between the α

and β curves. This is not the case when the transition is first-order (Fig. 20). The G vs. T diagram appropriate to the above considerations is shown in Fig. 22.

In summary, a second-order phase transition is a symmetry change which occurs with $\Delta S = \Delta V = 0$ at the transition point. It occurs during a continuous change in structure of the displacive or order-disorder type. The symmetry change must occur at a point since it is not possible for a structure to have a fraction of a symmetry operation. The point of loss or gain of symmetry is the point at which η goes to zero and it is the point at which the G vs T area (Fig. 22) converges to a line (volume becomes a surface in G-T-P space, etc.). At this point α and β do not coexist, but rather they become indistinguishable. At this point there is a discontinuity in C_p which will be discussed further in a later section.

It is perhaps worthwhile to note that there undoubtedly exist cases of first-order phase transitions which appear to be second-order. Such cases arise when the two G vs T curves of Fig. 19 correspond to nearly the same structure in the neighborhood of T_t. A fundamental distinction between first- and second-order processes, namely the coexistence of two phases at equilibrium in the first-order case, is then not useful because fluctuations in the experimental variables in either space or time will obscure the distinction. The result is that while it is possible to state that observed coexistence at equilibrium implies a first-order process, the apparent absence of such coexistence does not require a second-order process. Furthermore, it is not possible to say, simply because η decreases extremely rapidly to zero at T_t that a process is first-order because such rapid changes are allowed in a second-order process. It is not possible to devise an experiment which differentiates between an arbitrarily steep slope and a

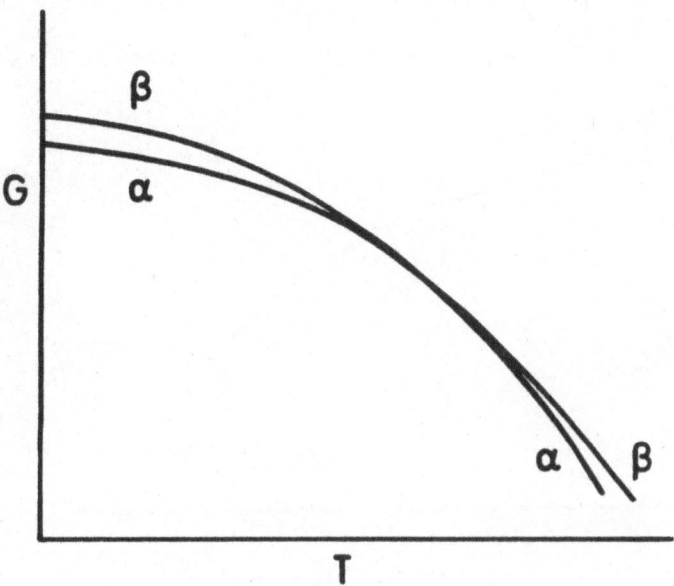

Fig. 21. G vs T for the second-order case with an unresolved problem (see text).

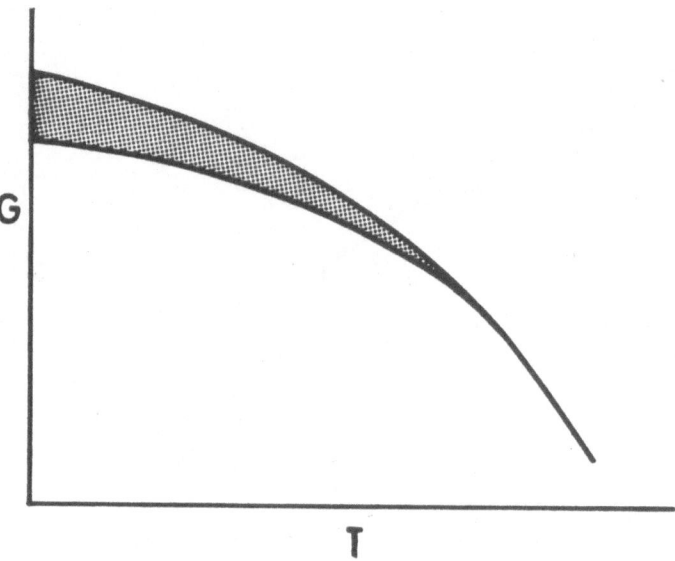

Fig. 22. G vs T for a second-order phase transition. The low-temperature phase is assumed to be the phase with lower symmetry.

discontinuity. Therefore in cases where coexistence at equilibrium cannot be demonstrated we are in need of additional criteria for the determination of the order of a transition and we turn to the theory of Landau (1).

III. LANDAU THEORY (WITHOUT SYMMETRY)

An essential feature of the Landau theory is the consideration of the behavior of G in the region of $\eta \neq \eta^{eq}$. This consideration is accomplished through the expansion of G in a Taylor's series in η for the cases of first- or second-order transitions:

$$G = G^\circ + \alpha\eta + A\eta^2 + B\eta^3 + C\eta^4 + \dots.$$

where G°, α, A, B, .. are evaluated at $\eta = 0$. It is customary, as was originally done by Landau, to terminate the expansion at the fourth-order term (although at one point in a later section it will be useful to consider the expansion to the η^6 term). It should be remembered that, as with G itself, those coefficients which need not vanish (as discussed below) are functions of thermodynamic state, e.g., of T, P, X.

First we note that the coefficient α must vanish because $\alpha = \left.\frac{\partial G}{\partial \eta}\right|_{T_t}$

and G must be at a minimum at $\eta = 0$ if $\eta = 0$ is to be stable. Thus if the thermodynamic states at which $\eta = 0$ is a stable phase are considered $\alpha(T,P,X) = 0$ and, since the transition points are included in these states, $\alpha = 0$ at the transition points. Furthermore stability of the $\eta = 0$ phase means G vs η is concave upwards at $\eta = 0$ and thus that

$$\left.\frac{\partial^2 G}{\partial \eta^2}\right]_{\eta=0} = A \geq 0 \quad.$$

Thus we have

$$G = G^\circ + A\eta^2 + B\eta^3 + C\eta^4$$

with $A \geq 0$. Clearly $C > 0$ or else G goes towards minus infinity and for large values of η would decrease G without bound i.e., a distortion catastrophy would result.

We can plot $G-G°$ vs η and obtain the plot of Fig. 23 for $B > 0$ (if $B < 0$ we simply reflect the plot through the vertical axis and all considerations follow). Note that a phase transition can occur when $G-G° = 0$, which will be shown to occur when $A(T,P,X)$ becomes sufficiently small relative to $B^2/4C$ as follows. Solving the expansion for the nonzero roots of $G-G° = 0$ yields

$$A\eta^2 + B\eta^3 + C\eta^4 = 0$$

or

$$A + B\eta + C\eta^2 = 0$$

and thus

$$\eta = \frac{-B \pm \sqrt{B^2 - 4AC}}{2C}$$

which has no real roots if $4AC > B^2$ (the situation of Fig. 23), which has one real root if $B^2 = 4AC$ (the situation of Fig. 24) and otherwise has two real roots. The case of $4AC > B^2$ thus corresponds to an absolute minimum at $\eta = 0$ (stable) and a relative minimum at $\eta \neq 0$ (metastable) and the case $4AC < B^2$ reverses the stability and metastability. The case of Fig. 24 (i.e., $B^2 = 4AC$) is the case of $\eta^{eq} = 0$ and $\eta^{eq} = -B/2C$ and two phases coexist in

equilibrium. This is the case of a first-order phase transition. Not all first-order phase transitions can be described in this way, for it is necessary for this description to be valid that the two phases be conceptually continuously related in structure. We note however, that two phases which are so related might necessarily transform via a first-order transition.

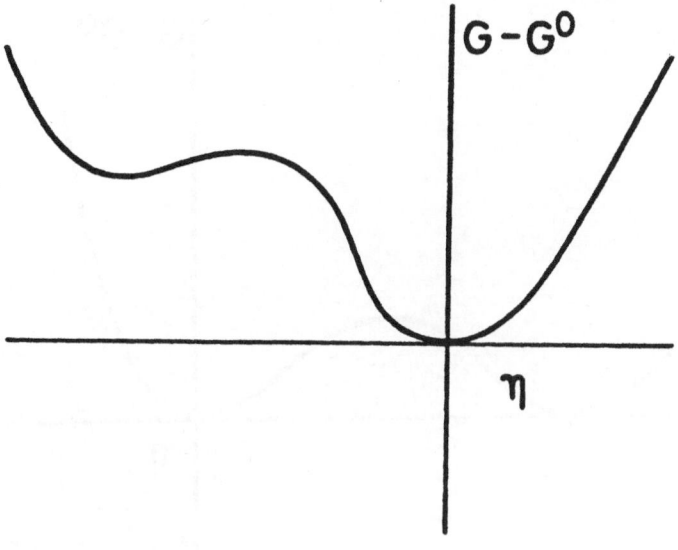

Fig. 23. G-G° vs η for A, B, C > 0.

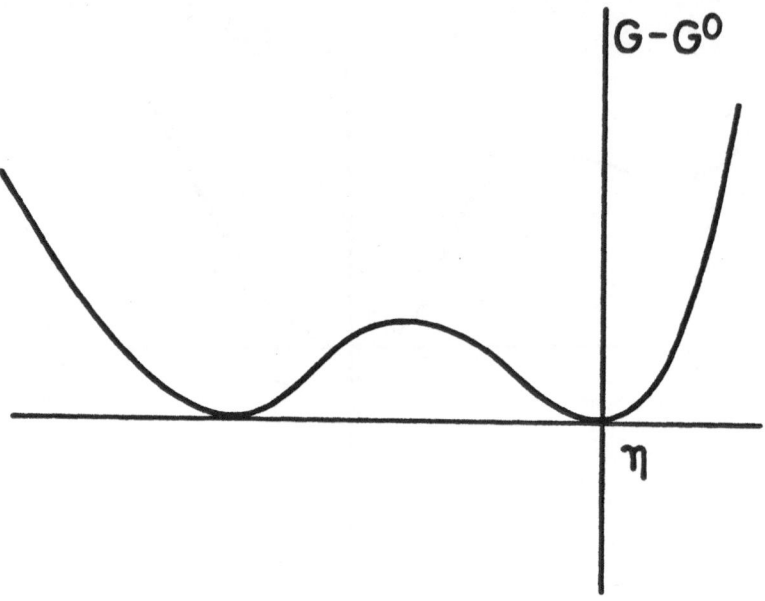

Fig. 24. $G-G°$ vs η for $B^2 = 4AC$.

Next we inquire into the behavior of G-G° if B ≡ 0, i.e., if B must vanish by symmetry (a condition which will be discussed in a later section), and is therefore zero at all T,P,X points. In this case

$$G = G° + An^2 + Cn^4$$

and we have the two possibilities shown in Fig. 25a.

In this case if A > 0 there is a single minimum at n^{eq} = 0 and if A < 0 there are two minima at n^{eq} ≠ 0 (recall that ± n correspond to equivalent distortions). Thus if A(T,P,X) goes through zero with changing thermodynamic state n^{eq} will in this case continuously change from zero to nonzero values (see Fig. 25b). This is the condition for a second-order phase transition and we see from the above it can occur only if B ≡ 0. This condition will be discussed from the point of view of symmetry in later sections.

The occurrence of a minimum of n^{eq} ≠ 0 for the B ≡ 0 case means

$$\frac{\partial G}{\partial n} = 2An_{eq} + 4Cn_{eq}^3 = 0$$

and thus

$$n_{eq}^2 = -\frac{A}{2C}.$$

Substitution into $G = G° + An^2 + Cn^4$ yields

$$G^{eq} = G° - \frac{A^2}{4C}.$$

Now, if A(T,P,X) is to vanish it must be a function of state, e.g., of T, and we can expand A in a series in T and drop all terms except the first in the neighborhood of T_t, i.e., take

$$A = \alpha (T-T_t)$$

with α > 0 and obtain

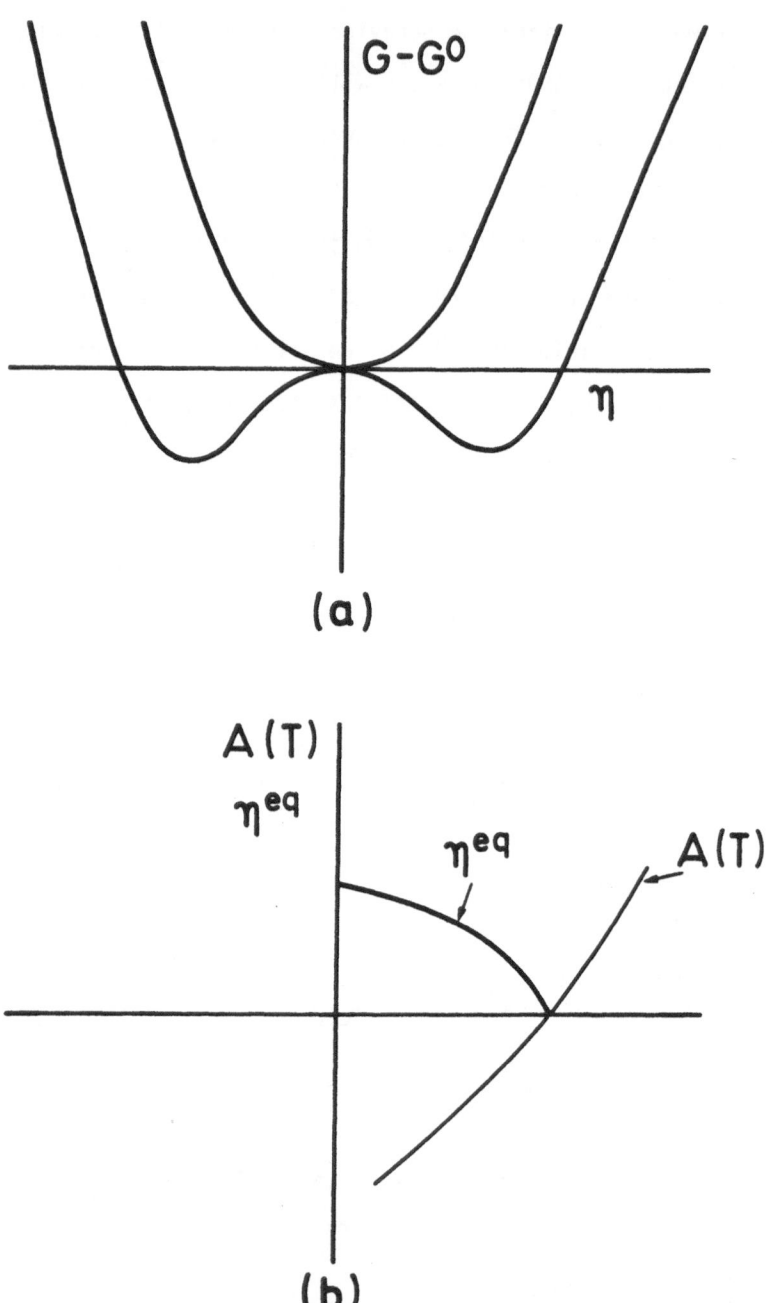

Fig. 25. Diagrams appropriate to $B \equiv 0$ and $A(T)$ changes sign. (a) $G-G^0$ vs η
$A>0$ yields single minimum and $A<0$ double minima at $\pm\eta^{eq} \neq 0$ (b) A and η^{eq} vs
T showing $A>0 \Rightarrow \eta^{eq} = 0$ and $A<0 \Rightarrow \eta^{eq} \neq 0$.

$$G^{eq} = G^\circ - \frac{\alpha^2 (T-T_t)^2}{4C} \, .$$

By thermodynamics $\left(\frac{\partial G}{\partial T} \right)_p = -S$ and thus

$$S^{eq} = S^\circ + \frac{\alpha^2 (T-T_t)}{2C}$$

from which, as desired, $\Delta S = 0$ at $T = T_t$.
Furthermore $\left(\frac{\partial S}{\partial T} \right)_p = \frac{C_p}{T}$ and thus

$$\frac{C_p^{eq}}{T} = \frac{C_p^\circ}{T} + \frac{\alpha^2}{2C}$$

and

$$\lim_{T \to T_t} \Delta C_p = \frac{\alpha^2 T_t}{2C} \, ,$$

and there is a discontinuity in C_p at the transition temperature.

Note that taking $A = \alpha (T-T_t)$ implies that the symmetrical ($\eta = 0$) form is the high temperature form. It is a fact that this is usually, if not always, the case for a second-order phase transition. If $A = \alpha (T_t-T)$ the other case (symmetrical form stable at $T < T_t$) would occur. In this case

$$S^{eq} = S^\circ - \frac{\alpha^2 (T_t-T)}{2C}$$

and if we consider $T > T_t$ we find $S^{eq} > S^\circ$. This corresponds to the physically awkward situation of a larger entropy for the ordered or distorted solid than for the $\eta = 0$ cases and therefore, because of the relationship $S = - \left(\frac{\partial G}{\partial T} \right)_p$, to a G vs T diagram as shown in Fig. 26. It follows that the observation that Fig. 22 is more frequently (or always) and Fig. 26 less frequently (or never) correct is correlated with the physically reasonable correlation of the greater entropy with the symmetrical form.

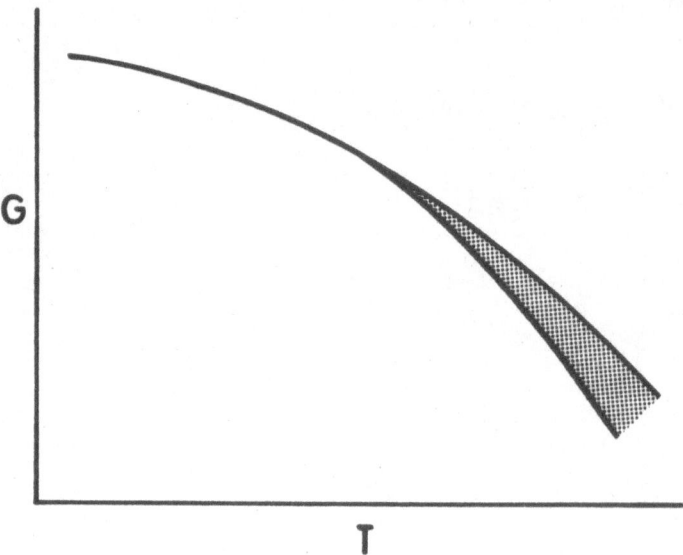

Fig. 26. G vs T for A = α (T_t-T) i.e., low-symmetry form stable at high temperature.

When Fig. 22 is correct the transition from unstable $\eta = 0$ to stable $\eta \neq 0$ at $T < T_t$ corresponds to a decrease in entropy whereas if Fig. 26 were correct the transition from unstable $\eta = 0$ to stable $\eta \neq 0$ at $T > T_t$ would occur with an increase in entropy. Such a change as implied by Fig. 26 would not correspond to the change in configurational entropy in CuZn or the expected change in the phonon density of states for VS (to be discussed later). In fact, for both CuZn and VS the high temperature forms are the symmetrical ones.

IV. LANDAU THEORY, WITH CONSIDERATION OF SYMMETRY, APPLIED TO THE NiAs-TYPE
    TO MnP-TYPE PHASE TRANSITION

The first condition of Landau is: in order that two crystalline structures be related by a second-order phase transition the space groups of the structures must be in the relation of a group and a subgroup. This follows from the continuity of the structure change -- symmetry elements can be destroyed at a state point by infinitesimal changes of occupancy or position, but simultaneous formation of symmetry operations is not possible because the creation of symmetry elements requires definite movement of atoms which cannot be accomplished by changes which are arbitrarily small in magnitude. For example the vectors of Fig. 18 destroy the hexagonal symmetry and maintain the vertical mirror plane to which they are parallel no matter what their magnitude. However to simultaneously create new symmetry elements in the structure would require specific, concerted movements, and could in no case be accomplished by vectors of length arbitrarily close to zero.

Therefore all of the symmetry elements in the low symmetry structure are present in the high symmetry structure and since, by presumption, both the high and low symmetry structures have symmetry operations which necessarily form groups, it follows that the symmetry operations of the low symmetry structure are a subgroup of those of the high symmetry structure. We note in passing that use of the words high and low in reference to the symmetries of the structures depends upon this relationship, for only if two groups are in the relationship of a group and a subgroup can it be meaningfully said that one structure is "more symmetrical" than the other.

In the case of the phase transition from NiAs ($P6_3/mmc$) to MnP (Pcmn), the symmetry operations are in a group-subgroup relation. This can be seen in a number of ways: by consulting the tables of Wondratschek (8) on groups and subgroups, by comparing figures in the International Tables (9) for Crystallography, Vol. 1 or by a detailed comparison of the symmetry operations of the two groups.

The symmetry elements so identified in Column 3 of Table 7 are present in both Pcmn and $P6_3/mmc$ and the others are lost in the transition (cf. Fig. 18 with the origin at the metal atom position in the hexagonal structure).

From the table it is clear that the symmetry operations of $P6_3/mmc$ fall into two categories: those that take x into itself or its negative and those that do not. Those that take x into ± x have β parts in $g_0(\vec{a}/2)$. If we examine the third column of the table we find that symmetry operations of the distorted structure are only among those that carry x into itself, or take x into -x <u>and</u> involve translation by $\vec{c}/2$. These are $\{\varepsilon|000\}$, $\{C_{2z} | 00^1/_2\}$ $\{\sigma_{2x+y} | 00^1/_2\}$ and $\sigma_y | 000\}$. The symmetry operations that carry x into -x <u>or</u> involve translation by $\vec{c}/2$ are not present in the distorted structure as they stand in the table, however they are present if translation by $\vec{a}$ is added (e.g., $\{i | 000\}$ is lost but $\{i | 100\}$ is not), and $\{C_{2z} | 10^1/_2\}$, $\{C_{2y} | 100\}$, $\{i | 100\}$ and $\{\sigma_z | 10^1/_2\}$ are found in the distorted

TABLE 7.

Symmetry Operations of P6$_3$/mmc	Transformation of x,y,z	Present in distorted structure?	cos(2πz)sin(πx)
{ε | 000}	x,y,z	yes	+1
{C$_{6z}$ | 001/2}	x-y,x,z+1/2		
{C$_{6z}^2$ | 000}	ȳ,x-y,z		
{C$_{2z}$ | 001/2}	x̄,ȳ,z+1/2	yes	+1
{C$_{6z}^4$ | 000}	y-x,x̄,z		
{C$_{6z}^5$ | 001/2}	y,y-x,z+1/2		
{C$_{2y}$ | 000}	x̄,y-x,z̄		-1
{C$_{2(x-y)}$ | 001/2}	ȳ,x̄,z̄+1/2		
{C$_{2x}$ | 000}	x-y,ȳ,z̄		
{C$_{2(x+y)}^2$ | 001/2}	x,x-y,z̄+1/2		-1
{C$_{2(x+y)}$ | 000}	y,x,z̄		
{C$_{2(x+2y)}$ | 001/2}	y-x,y,z̄+1/2		
{i | 000}	x̄,ȳ,z̄		-1
{C$_{6z}$ | 001/2}	y-x,x̄,z̄+1/2		
{C$_{6z}^{-2}$ | 000}	y,y-x,z̄		
{σ$_z$ | 001/2}	x,y,z̄+1/2		-1
{C$_{6z}^{-4}$ | 000}	x-y,x,z̄		
{C$_{6z}^{-5}$ | 001/2}	ȳ,x-y,z̄+1/2		
{σ$_y$ | 000}	x,x-y,z	yes	+1

TABLE 7 - continued

Symmetry Operations of P6$_3$/mmc	Transformation of x,y,z	Present in distorted structure?	cos(2πz)sin(πx)
{σ$_{x+y}$ | 001/2}	y,x,z+1/2		
{σ$_x$ | 000}	y-x,y,z		
{σ$_{2x+y}$ | 001/2}	x̄,y-x,z+1/2	yes	+1
{σ$_{x-y}$ | 000}	ȳ,x̄,z		
{σ$_{x+2y}$ | 001/2}	x-y,ȳ,z+1/2		

structure. These symmetry operations that are found in the distorted structure form the space group Pcmn (minus the pure translations, which are also present in the distorted structure as is clear from Fig. 18), and thus the space group of the distortion is Pcmn and it is a subgroup of P6$_3$/mmc.

The second condition of Landau theory is that the distortion should correspond to a single irreducible representation of the space group of higher symmetry. This is so because, as will be discussed later, the coefficients in the expansion of G in powers of η (the A,B,C,..) can be associated with a single irreducible representation. That is, a coefficient A corresponding to one irreducible representation and another, A', corresponding to another, are not equal by symmetry and therefore will be different functions of thermodynamic state and will be equal at most at isolated state points. Accordingly two A functions will not vanish at a succession of state points and behavior such as shown in Fig. 19a will not be observed. Thus, in general, the behavior of Fig. 19a corresponds to a sign change of one A, and the distortion is the one appropriate to the irreducible representation to which that A corresponds. This point will be made explicit in later sections.

An examination of the 4th column of Table 7 shows the behavior of the function $\phi_1 = \cos(2\pi z)\sin(\pi x)$ under the operations that take x into ± x. It follows that ϕ_1 is a basis function for a small irr. rep. at $\vec{a}/2$. All of the operations that remain in the distorted structure take ϕ_1 into ϕ_1, and those that are lost take ϕ_1 into $\pm\phi_2$ (= $\cos(2\pi z)\sin(\pi y)$, $\pm\phi_3$ (= $\cos(2\pi z)\sin \pi(x-y)$) or - ϕ_1. Accordingly the distortion corresponds to a single irreducible representation of P6$_3$/mmc, and ϕ_1 is one of the basis functions for this representation. Note that if x in ϕ_1 is replaced by x+1 (translation by $\vec{a}$) then the sign of the function is changed, and thus those operations which take ϕ_1 into $-\phi_1$ take ϕ_1 into ϕ_1 when $\vec{a}$ is added to the translational part.

The significance of the correspondence of the distortion with the irreducible representation is, as will be shown, that there exists an electron density function $\Delta\rho$, which yields $\rho^{low \ symmetry}$ when added to $\rho^{high \ symmetry}$ and which is a basis function for a single irreducible representation which is symmetric with respect to symmetry operations of the high symmetry group which remain and antisymmetric with respect to those that are lost. Thus $\Delta\rho$ destroys and maintains the appropriate symmetry operations when added to ρ^{high}. The significance of the functions ϕ_2 and ϕ_3 is that they are rotated (by 120° and 240°, respectively) ϕ_1's which correspond to alternative Pcmn distortions in the symmetrically equivalent directions of the hexagonal lattice.

The third condition of Landau is that no third-order combination of the basis functions be invariant under the symmetry operations of the group. As will be shown below this assures that B≡0. It is possible to consider the operation of the symmetry operations on the set of functions ϕ_1, ϕ_2, ϕ_3, for example $\{C_{6z}^5 \mid 001/2\}$ which takes x,y,z into y, y-x, z+1/2 and thus ϕ_1 into

$-\phi_2$, ϕ_2 into ϕ_3 and ϕ_3 into $-\phi_1$, i.e.

$$\begin{pmatrix} 0 & \bar{1} & 0 \\ 0 & 0 & 1 \\ \bar{1} & 0 & 0 \end{pmatrix} \begin{pmatrix} \phi_1 \\ \phi_2 \\ \phi_3 \end{pmatrix} = \begin{pmatrix} -\phi_2 \\ \phi_3 \\ -\phi_1 \end{pmatrix}.$$

In this fashion we can generate a set of 3x3 matrices which multiply like the symmetry operations of P6$_3$/mmc, i.e., a 3x3 representation (which is irreducible) of the space group with the basis functions ϕ_1, ϕ_2 and ϕ_3. Although the ϕ's are useful for the consideration of symmetry, they are not electron density functions, however such functions with the same symmetries could be obtained (i.e., $\Delta\rho$'s which are also basis functions for the irreducible representation) for example by calculating ρ for NiAs-type and for MnP-type and taking the difference with the 3 different orientations of the MnP-type structure.

A general distortion corresponding to this irr. rep. is given by

$$\rho^{high} - \rho^{low} = C_1(\Delta\rho_1) + C_2(\Delta\rho_2) + C_3(\Delta\rho_3),$$

and the MnP-type distortion that doubles the a axis corresponds to $C_1 \neq 0$ and $C_2 = C_3 = 0$.

Since $\rho^{low} \rightarrow \rho^{high}$ as the C_i's $\rightarrow 0$ it is possible to expand G for the general distortion about G° for the symmetrical structure:

$$G = G° + Af^{\langle 2 \rangle}_{(C_i)} + Bf^{\langle 3 \rangle}_{(C_i)} + Cf^{\langle 4 \rangle}_{(C_i)}$$

where, as previously discussed, there is no first-order term, A is the coefficient of the second-order terms corresponding to this irr. rep. (and is the same for C_1^2, C_2^2 and C_3^2 because distortions corresponding to $\Delta\rho_1$, $\Delta\rho_2$ and $\Delta\rho_3$ are equivalent by symmetry) and $f^{\langle 2 \rangle}_{(C_i)}$ is the second-order combination of the C_i's that is invariant under the symmetry operations of P6$_3$/mmc, etc.

This point concerning invariance requires further amplification. Consider $\Delta\rho = C_1(\Delta\rho_1) + C_2(\Delta\rho_2) + C_3(\Delta\rho_3)$ and the effect of any of the symmetry operations of $P6_3/mmc$ upon $\Delta\rho$. For example $\{i \mid 000\}$ takes $C_1(\Delta\rho_1) + C_2(\Delta\rho_2) + C_3(\Delta\rho_3)$ into $-C_1(\Delta\rho_1) -C_2(\Delta\rho_2) -C_3(\Delta\rho_3)$, i.e., we can write

$$
\begin{pmatrix} \bar{1} & 0 & 0 \\ 0 & \bar{1} & 0 \\ 0 & 0 & \bar{1} \end{pmatrix} \begin{pmatrix} C_1 \\ C_2 \\ C_3 \end{pmatrix} = \begin{pmatrix} -C_1 \\ -C_2 \\ -C_3 \end{pmatrix}
$$

and similarly we can find 3x3 matrices (multiplying the C_i column matrix) for all other operations. Thus the C_i's form a basis for the same irr. rep. as do the ϕ_i's. The sum $C_1^2 + C_2^2 + C_3^2$ is invariant under all symmetry operations of $P6_3/mmc$ since such operations only permute and change the signs of the C_i's. Since G for any symmetrically equivalent distortion of $P6_3/mmc$ must be the same, G is invariant under symmetry operations and so too must be the terms of G. Accordingly $C_1^2 + C_2^2 + C_3^2$ is a possible second-order term in this expansion, however $C_1C_2 + C_2C_3 + C_1C_3$ is not because under some symmetry operations of $P6_3/mmc$ this term is not invariant. We find that there is only one invariant of second-order, namely ΣC_i^2, and in fact such terms are generally the only second order invariants as shown by a theorem of group theory.

Attempts to find third-order invariants fail e.g., $C_1C_2C_3$ goes into $-C_1C_2C_3$ under a number of operations of $P6_3/mmc$ and also $C_1^3 + C_2^3 + C_3^3$ varies etc. Thus there is no third-order invariant ($f_{(C_i)}^{<3>} \equiv 0$) and no third order term. As regards fourth-order invariants examination shows that $C_1^4 + C_2^4 + C_3^4$ and $C_1^2C_2^2 + C_2^2C_3^2 + C_1^2C_3^2$ are both invariant (10) and are independent of each other, i.e., no symmetry operation carries one into the other. Thus

$$f^{\langle 4 \rangle}_{(C_i)} = C_1^4 + C_2^4 + C_3^4 + K \left(C_1^2 C_2^2 + C_2^2 C_3^2 + C_1^2 C_3^2 \right)$$

where $K = K(T,P,X)$.

At this point it is convenient to introduce a definition of γ_i:

$$\gamma_i^2 = C_i / \Sigma C_i^2$$

(i.e., the γ_i's are normalized C_i's in the sense that $\Sigma \gamma_i^2 = 1$) and the definition of $\eta = \Sigma C_i^2$. It follows that

$$G = G^\circ + A\eta^2 + C[\gamma_1^4 + \gamma_2^4 + \gamma_3^4 \ K(\gamma_1^2 \gamma_2^2 + \gamma_2^2 \gamma_3^2 + \gamma_1^2 \gamma_3^2)]\eta^4$$

or

$$G = G^\circ + A\eta^2 + C'\eta^4$$

where $C' = C[\gamma_1^4 + \gamma_2^4 + \gamma_3^4 + K(\gamma_1^2 \gamma_2^2 + \gamma_2^2 \gamma_3^2 \gamma_1^2 \gamma_3^2)]$, and we return to the form of the G vs. η expression used in the preceeding section.

The effort expended above to include symmetry information in the expansion of G has resulted in additional structural information which is contained in $G(\gamma_i)$. Namely, stable structures can result only when $G(\gamma_i)$ is at a minimum. We thus seek the minima of $G(\gamma_i)$ subject to the restraint $\Sigma \gamma_i^2 = 1$. This amounts to minimizing $\gamma_1^4 + \gamma_2^4 + \gamma_3^4 + k(\gamma_1^3 \gamma_2^2 + \gamma_2^1 \gamma_3^2 + \gamma_1^2 \gamma_3^2)$ subject to this restraint. This can be accomplished using Lagrange's method of undetermined multipliers. However since

$$\gamma_1^2 + \gamma_2^2 + \gamma_3^2 = 1$$

it follows that

$$\gamma_1^4 + \gamma_2^4 + \gamma_3^4 + 2(\gamma_1^2 \gamma_2^2 + \gamma_2^2 \gamma_3^2 + \gamma_1^2 \gamma_3^2) = 1$$

and thus that the function to be minimized can be written

$$1 + (K+2)(\gamma_1^2\gamma_2^2 + \gamma_2^2\gamma_3^2 + \gamma_1^2\gamma_3^2)$$

and we can see that this function is minimized if $\gamma_1^2\gamma_2^2 + \gamma_2^2\gamma_3^2 + \gamma_1^2\gamma_3^2$ is maximized when K-2 < 0 and if $\gamma_1^2\gamma_2^2 + \gamma_2^2\gamma_3^2 + \gamma_1^2\gamma_3^2$ is minimized when K-2 > 0, i.e., there are two solutions, one stable if K < 2, the other if K > 2. The smallest value that $\gamma_1^2\gamma_2^2 + \gamma_2^2\gamma_3^2 + \gamma_1^2\gamma_3^2$ can have is zero and it has this value if γ_1 = 1 and γ_2 = γ_3 = 0. This corresponds to the MnP-type distortion and we have now found that this is a stable solution.

A small amount of playing with numbers reveals that $\gamma_1^2\gamma_2^2 + \gamma_2^2\gamma_3^2 + \gamma_1^2\gamma_3^2$ is maximized if γ_1 = γ_2 = γ_3 = $1/\sqrt{3}$ and thus this is the stable solution when K < 2. The structure to which this solution corresponds can be revealed by considering the vectors corresponding to the distortions of $\Delta\rho_1$, $\Delta\rho_2$ and $\Delta\rho_3$ (as has been done for $\Delta\rho_1$ in Fig. 2) and adding the vectors. The result is a hexagonal structure with a and b doubled relative to the NiAs-type. This structure is known for the low-temperature form of NbS. The high-temperature form of NbS is MnP-type, and thus it seems likely that, at least in principle, a yet higher form (NiAs-type) exists and that K = 2 somewhere in the homogeneity range of NbS. In summary, we have shown that the Landau theory provides relationships between structures (e.g., between NbS (1.t.) and MnP-type) which correspond to possible stable solutions for the same irr. rep. of a space group, and indicates the existence of structure types that might be sought experimentally.

In closing this section it is important to remark that even if B≡0 a transition still need not be second-order. In order to consider this possibility it is necessary to consider terms to 6th order (5th and higher odd order invariants do not exist):

$$G = G° + A\eta^2 + C\eta^4 + E\eta^6 .$$

Now if A, E > 0 and C < 0 we have G-G° vs. η as shown in Fig. 27. The minima at $\eta \neq 0$ are reminiscent of that with B $\neq$ 0 and the behavior is similar, i.e., a first-order transition necessarily results.

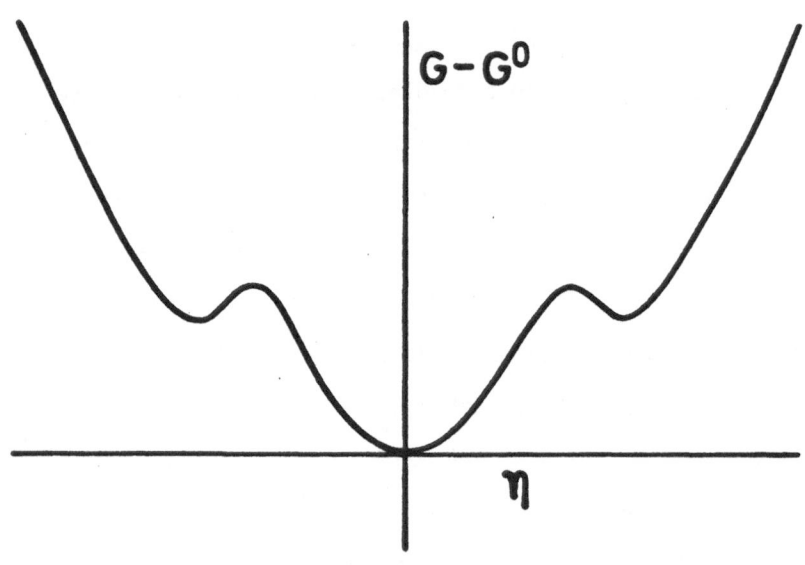

Fig. 27. G-G° vs. η with terms to η^6. B$\equiv$0, C<0, A,E>0.

V. GENERAL DEVELOPMENT OF THE LANDAU THEORY WITH CONSIDERATION OF SYMMETRY

The electron density ρ, of the low symmetry phase is expressed as a linear combination of basis functions for the irr. reps. of the space group of higher symmetry.

$$\rho = \sum_{\alpha} \sum_{i} C_i^{\alpha} (\Delta\rho_i)^{\alpha}$$

where α labels the irr. reps. and i the degenerate basis functions for a given irr. rep. That this may be done is a theorem of group theory, and the procedure whereby the coefficients would be found once the $(\Delta\rho_i)^{\alpha}$'s were known is called projection.

Among the irr. reps. there is always one corresponding to ρ° called the totally symmetric which has the symmetry of the high-symmetry space group. This term is removed from the sum

$$\rho = \rho^{\circ} + \sum_{\alpha}' \sum_{i} C_i^{\alpha} (\Delta\rho_i)^{\alpha} .$$

Since $\rho \to \rho^{\circ}$ as C_i's $\to 0$, G, the Gibbs free energy of a general distortion is expanded in the C_i's

$$G = G^{\circ} + A_1 n_1^2 + A_2 n_2^2 + \dots$$

where we make use of the fact that the only second-order invariant corresponding to each irr. rep. is $\sum C_i^2 = n_{\alpha}$. Now continuous distortion occurs when one of the A_{α}'s changes sign and, since the A_{α}'s are independent, with very high probability only one A_{α} will change sign at a given state point. Even if a point at which two A_{α}'s vanished did occur it would not be the case that two A_{α}'s would change sign at a succession of state points such as represented by the line of Fig. 19a. Accordingly we can say that a second-order phase transition that occurs at a succession of state points (T vs. X or T vs. P, etc.) will correspond to a single irr. rep. of the space group of higher symmetry.

Thus we consider only the terms for a single irr. rep. i.e.,

$$G = G° + An^2 + Bf(\gamma_i)^{<3>}n^3 + Cf(\gamma_i)^{<4>}n^4$$

to terms of fourth order. We have seen that in order for the transition to be second-order the coefficient of n^3 must vanish. Since symmetry operations carry

$$\rho = \rho° + \sum_i C_i \ (\Delta\rho_i)$$

into equivalent electron densities it follows that G must be invariant under symmetry operations (equivalent distortions have the same G). Such symmetry operations alter the coefficients of the basis functions, and thus one must examine whether $f(\gamma_i)^{<n>}$ exists and what its form is for various n's. For n=1 there exists no invariant except for the totally symmetric irr. rep. which would correspond to no change in symmetry. For n=2, $f(\gamma_i)^{<2>} = \Sigma\gamma_i^2$ is always the only invariant that need be considered. For n=3 there may or may not be invariants, and this must be investigated. It is necessary for a second-order phase transition to occur that there not be third-order invariants. For n=4, $(\Sigma\gamma_i^2)^2$ is always one invariant, however the number of independent 4th order invariants must be determined. If there is more than one then there is more than one possible distortion corresponding to the irr. rep. To determine the stable solutions it is necessary to minimize $G(\gamma_i)$ subject to the restraint $\Sigma\gamma_i^2 = 1$.

VI. LANDAU'S 4th CONDITION (11)

We have seen that A, the lead coefficient in G-G°, corresponds to a given irr. rep. and that the irr. reps. correspond to particular k vectors. It follows that A depends upon $\vec{k}$ and in fact can be expanded in $\delta\vec{k}$ about a given $\vec{k}$, i.e.,

$$A(\vec{k} + \delta\vec{k}) = A(\vec{k}) + \vec{\alpha} \cdot \delta\vec{k} + \ldots$$

where $\vec{\alpha}$ must be a vector and the first order term is a dot product because A is a scalar. In general $\vec{\alpha}(T,P,X)$ will exist and be nonzero and therefore some $\delta\vec{k}$ for which $\vec{\alpha} \cdot \delta\vec{k} < 0$ will necessarily exist, and $A(\vec{k}+\delta\vec{k}) < A(\vec{k})$ is assured. However if this is the case then a distortion at point k is not allowed since there exists a more stable point in the neighborhood of $\vec{k}$ (namely at $\vec{k}+\delta\vec{k}$) and thus the whole consideration given in the earlier sections to the vanishing of A is irrelevant since the distortion is to a different k point with a different $g_0(k)$ and a different set of irr. reps. The only way in which the treatment given above can be rescued is if $\vec{a}\equiv0$, i.e., if there can be no vector invariant for reasons of symmetry. That is, suppose that the symmetry of reciprocal space is such that $\delta\vec{k}$ and $\beta\delta\vec{k}$ must have the same A value (i.e., suppose β is in $g_0(\vec{k})$). Then

$$\vec{\alpha} \cdot \delta\vec{k} = \vec{\alpha} \cdot \beta\delta\vec{k}$$

or

$$\vec{\alpha} \cdot \delta\vec{k} = \beta^{-1}\vec{\alpha} \cdot \delta\vec{k}$$

for all β and $\delta\vec{k}$, i.e., $\vec{\alpha} = \beta^{-1}\vec{\alpha}$. This will be possible with $\vec{\alpha} \neq 0$ for some special cases, e.g., a single axis ($\vec{\alpha}$ lies along axis) or plane ($\vec{\alpha}$ lies in plane). However if β^{-1} is inversion, or if there are two β's such as an intersecting plane and axis or intersecting pair of axes, then the above can be true only if $\alpha \equiv 0$ and a second phase transition to $\vec{k}$ is possible. The fourth condition of Landau then requires that there be an inversion or a pair of intersecting axes or an intersecting axis and mirror in $g_o(k)$ in order that a second-order phase transition be possible. Note that the case of $P6_3/mmc$ at $k = \vec{a}^*/2$ meets this condition since i is included in D_{2h}.

However, an implicit assumption in the above discussion is that the unsymmetrical structure has space group symmetry, and thus that the transition is to a $\vec{k}$ point which corresponds to a given fixed periodicity (such as $k = \vec{a}^*/2$ which doubles the periodicity in the $\vec{a}$ direction). Suppose instead (11) that the transition is to a variable $\vec{k}$ (e.g., to $\vec{k} + \delta\vec{k}$ with $\delta\vec{k}$ variable). Then the transition can occur to a point at which $\vec{\alpha}(T,P,X)$ happens to be zero, and that point will change, for example with changing composition or pressure. Such a case is an incommensurate phase (since δk is not an integral submultiple of a reciprocal lattice vector) and the characteristic length and direction of the incommensurate portion of the structure varies with state. In this case the 4th condition, which is appropriate only for transitions to a given $\vec{k}$ point, is not applicable.

VII. ANOTHER EXAMPLE OF THE APPLICATION OF LANDAU THEORY: Fm3m-R$\bar{3}$m ORDER-DISORDER TRANSITION IN $Sc_{1-x}S$

Stoichiometric ScS has the NaCl-type structure. At high temperatures the compound becomes nonstoichiometric with Sc vacancies. When the sample has been cooled slowly the vacancies are preferentially located in alternate Sc containing planes along the [1,1,1] direction (Fig. 28). Fm3m symmetry elements that remain after ordering are the identity, the 3-fold rotations along [1,1,1], the two-folds perpendicular to [1,1,1], the inversion and the product of all the aforementioned with the inversion, i.e, the operations

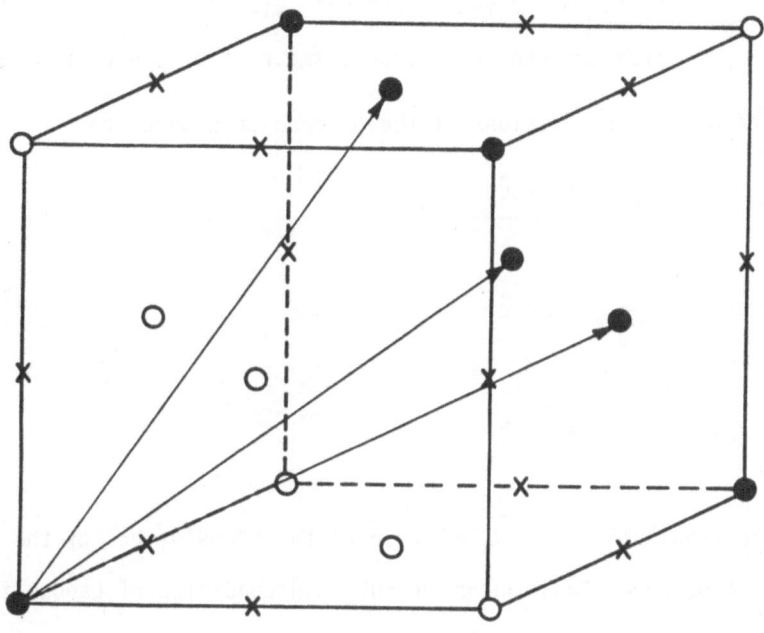

Fig. 28. R3̄m superstructure of the NaCl-type structure.

that remain are ε, C_3, C_3^2, $C_{2(y-x)}$, $C_{2(z-y)}$, $C_{2(z-x)}$, i, C_3^-, C_3^{-5}, $\sigma_{(y-x)}$, $\sigma_{(z-y)}$, $\sigma_{(z-x)}$, which are the rotational symmetry elements of the space group R$\bar{3}$m. The primitive translations of the ordered structure are

$$\vec{a}_{rh} = \vec{a}_c + \frac{\vec{b}_c + \vec{c}_c}{2}$$

$$\vec{b}_{rh} = \vec{b}_c + \frac{\vec{a}_c + \vec{c}_c}{2}$$

and

$$\vec{c}_{rh} = \vec{c}_c + \frac{\vec{a}_c + \vec{b}_c}{2} \quad ,$$

and thus the translations are a subgroup of the translations of the cubic, disordered structure. Thus the group-subgroup condition of Landau is met by this order-disorder transition.

Next we examine whether the distortion corresponds to a single irr. rep. Since all of the translational symmetry operations that are lost obey

$$(\vec{a}^* + \vec{b}^* + \vec{c}^*) \cdot \vec{T} = n/2, \; n = \text{odd integer}$$

and all that remain obey

$$(\vec{a}^* + \vec{b}^* + \vec{c}^*) \cdot \vec{T} = n, \; n = \text{integer}$$

it follows that the distortion "corresponds to" the $\vec{k}$ vector $(\vec{a}^* + \vec{b}^* + \vec{c}^*)/2$, i.e., $\cos(2\pi(\vec{a}^* + \vec{b}^* + \vec{c}^*)/2 \cdot \vec{r}) = \cos \pi(x+y+z)$ is a basis function for the irr. rep. that is symmetric with respect to translations that remain and antisymmetric with respect to those that are lost. Furthermore all of the rotational operations listed above are in the point group of $(\vec{a}^* + \vec{b}^* + \vec{c}^*)/2$, thus the ordering corresponds to the totally symmetric small representation of Fm3m at $\vec{k} = {}^1/_2, {}^1/_2, {}^1/_2$, and the symmetry change does correspond to a single irr. rep.

In order to consider the invariants of various order we must determine what $\vec{k}$ vectors are in the star. The symmetry operations of Fm3m carry $1/2$, $1/2$, $1/2$ into $-1/2$, $1/2$, $1/2$; $1/2$, $-1/2$, $1/2$; and $1/2$, $1/2$, $-1/2$ as well as the four vectors that result from these by inversion. Each $\vec{k}$ vector is related to its inverse by 1,1,1, and thus $\vec{k} = i\vec{k} + \bar{\vec{k}}$. Note that although $1/2$, $1/2$, $1/2$ and $-1/2$, $1/2$, $1/2$ differ by 1,0,0 they are not the same modulo a reciprocal lattice vector because 1,0,0 is not a reciprocal lattice vector in the case of a face-centered real lattice. There are four basis functions which form a basis for the irr. rep. to which the transition corresponds:

$$\Phi_1 = \cos \pi \ (x+y+z)$$

$$\Phi_2 = \cos \pi \ (-x+y+z)$$

$$\Phi_3 = \cos \pi \ (x-y+z)$$

$$\Phi_4 = \cos \pi \ (x+y-z),$$

and the rotational symmetry operations carry these functions into each other without sign change, while the translations carry them into themselves (if the translations remain) or into their negative (if the translations are lost).

It follows that in the expansion for G:

$$G = G^\circ + An^2 + \overset{\langle 3 \rangle}{Bf(\gamma_i)n^3} + \overset{\langle 4 \rangle}{Cf(\gamma_i)n^4} + \ldots$$

based upon a generalized electron density corresponding to this irr. rep.:

$$\rho^{\text{distorted}} = \rho^{\text{sym}} + [\gamma_1\Delta\rho_1 + \gamma_2\Delta\rho_2 + \gamma_3\Delta\rho_3 + \gamma_4\Delta\rho_4]\ n \ ,$$

there are no third-order invariants and the fourth-order term looks like:

$$C_1 \Sigma \gamma_2^4 + C_2 \sum_{i \neq j} \gamma_i^2 \gamma_j^2 + C_3 \gamma_1 \gamma_2 \gamma_3 \gamma_4 \; .$$

The stable distortions will correspond to minima of this function subject to the restraint $\Sigma \gamma_i^2 = 1$. Employing Lagrange's method of undetermined

multipliers with λ as the multiplier and minimization with respect to γ_1, γ_2,

γ_3 and γ_4 yields:

$$4C_1 \gamma_1^3 + 2C_2 \gamma_1 (\gamma_2^2 + \gamma_2^2 + \gamma_4^2) + C_3 \gamma_2 \gamma_3 \gamma_4 + 2\lambda \gamma_1 = 0$$

$$4C_1 \gamma_2^3 + 2C_2 \gamma_2 (\gamma_1^2 + \gamma_3^2 + \gamma_4^2) + C_3 \gamma_1 \gamma_3 \gamma_4 + 2\lambda \gamma_2 = 0$$

$$4C_1 \gamma_3^2 + 2C_2 \gamma_3 (\gamma_1^2 + \gamma_2^2 + \gamma_4^2) + C_3 \gamma_1 \gamma_2 \gamma_4 + 2\lambda \gamma_3 = 0$$

$$4C_1 \gamma_4^2 + 2C_2 \gamma_4 (\gamma_1^2 + \gamma_2^2 + \gamma_3^2) + C_3 \gamma_1 \gamma_2 \gamma_3 + 2\lambda \gamma_4 = 0$$

for which there are three types of solutions, $\gamma_1=1$, $\gamma_2=\gamma_2=\gamma_4=0$ (the R$\bar{3}$m ordering observed), $\gamma_1=\gamma_2=\gamma_3=\gamma_4=1$ (a f.c.c. ordering with a = 2 a_{NaCl}), and $\gamma_1=\gamma_2=1/\sqrt{2}$, $\gamma_3= \gamma_4=0$ (an orthorhombic ordering with a = $\sqrt{2}\, a_{NaCl}$), b = 2 a_{NaCl} and c = $a_{NaCl}/\sqrt{2}$.

VIII. APPLICATION OF LANDAU THEORY TO A TRANSITION TO A STRUCTURE WITH A
       MIXED ORDER-DISORDER AND DISTORTED STRUCTURE (6)

       Finally, consider the case of a tendency towards order (as a
decreased configuration entropy) in which the ordering can be accommodated
only by a distortion (e.g., the ordering of the ions requires a shearing of
the structure). For example (6) NaNO$_2$ at high temperatures has the symmetry
mmm with the NO$_2^-$ groups oriented randomly in the $\pm \vec{b}$ directions (Fig. 29a).
When the temperature is lowered they tend to order. At low temperatures the
structure adopted is that in which all NO$_2^-$ ions have their dipoles aligned
(ferroelectric structure, Fig. 29b). The symmetry of this structure is m2m.

However when the ordering is continuous there occur domains in which the ferroelectric structure is found, and boundary regions between ferroelectric domains in which the NO_2^- dipoles are aligned antiferroelectrically (e.g., in one direction for $z=0$ and in the opposite direction at $z=\frac{1}{2}$, Fig. 29c). Thus the structure appears to X-rays as a mixture of the ferroelectric and antiferroelectric structures during the continuous ordering from the paraelectric to the ferroelectric structure. The antiferroelectric structure requires a movement of the layer at $z=\frac{1}{2}$ such that the symmetry distorts to $2_z/m$. Thus the continuous transition from mmm toward m2m proceeds through an intermediate phase which is a mixture of domains of m2m and $2_z/m$.

At point Γ, $2_z/m$ and m2m correspond to different irr. reps. of mmm (Γ_6 and Γ_7, see Table 8). Therefore such a continuous distortion cannot occur at Γ. On the other hand a distortion to an incommensurate phase with $\vec{\delta k} = \delta \vec{a}^*$ can occur because the group of the wave vector in this case is C_{2x}, m_y and m_z and both Γ_6 and Γ_7 of mmm are compatible with Γ_3 of 2 mm (Table 8). Thus the process is possible as a second-order transition at $\delta \vec{a}^*$ allowing the coupling of the two processes and resulting in an incommensurate structure.

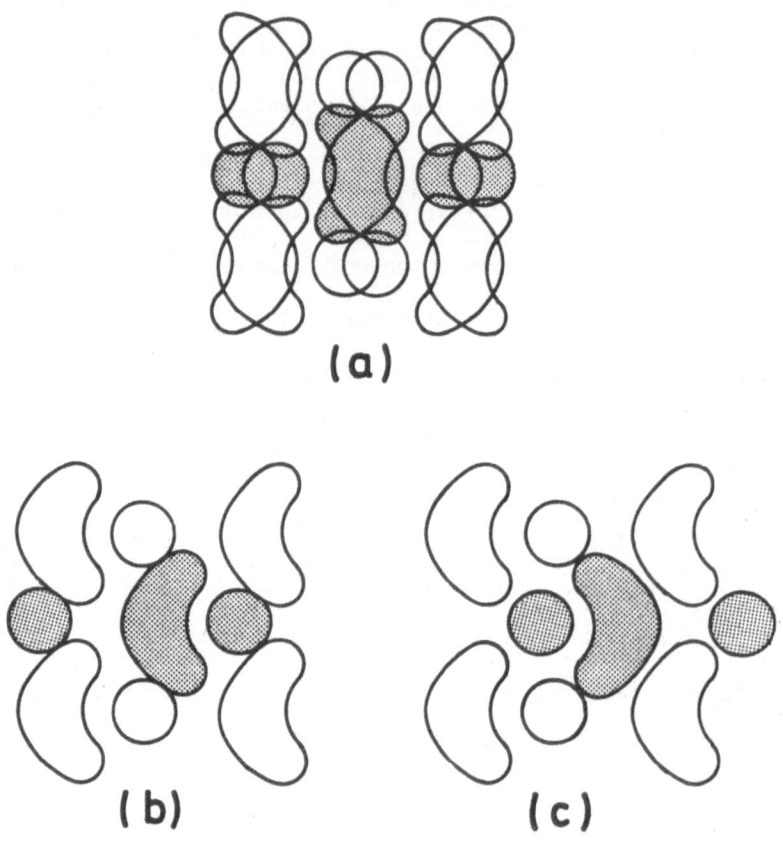

Fig. 29. Projections of the high-temperature (m2m) form (a), the low-temperature ferroelectric form (m2m) (b) and a hypothetical antiferroelectric form ($2_1/m$) (c) of $NaNO_2$. The kidney shaped objects represent NO_2, the circles represent Na. The shaded objects are at $z = 1/2$, the unshaded ones at $z = 0$.

TABLE 8. Irr. reps. of D2h and C2h.

mmm	ϵ	C_{2x}	C_{2y}	C_{2z}	i	σ_x	σ_y	σ_z
Γ_1	1	1	1	1	1	1	1	1
Γ_2	1	1	1	1	-1	-1	-1	-1
Γ_3	1	1	-1	-1	1	1	-1	-1
Γ_4	1	1	-1	-1	-1	-1	1	1
Γ_5	1	-1	1	-1	1	-1	1	-1
Γ_6	1	-1	1	-1	-1	1	-1	1
Γ_7	1	-1	-1	1	1	-1	-1	1
Γ_8	1	-1	-1	1	-1	1	1	-1

2mmm	ϵ	C_{2x}	σ_y	σ_z
Γ_1	1	1	1	1
Γ_2	1	-1	1	-1
Γ_3	1	-1	-1	1
Γ_4	1	1	-1	-1

Appendix 1

Rotation by 60° in the hexagonal lattice carries x,y into x-y,x (see Fig. 10), i.e.,

$$C_6 = \begin{pmatrix} 1 & \bar{1} \\ 1 & 0 \end{pmatrix}$$

and thus $C_3 = C_6^2 = \begin{pmatrix} 1 & \bar{1} \\ 1 & 0 \end{pmatrix} \begin{pmatrix} 1 & \bar{1} \\ 1 & 0 \end{pmatrix} = \begin{pmatrix} 0 & \bar{1} \\ 1 & \bar{1} \end{pmatrix}$

Accordingly we have

$$\begin{pmatrix} 1 & 0 & t_1 \\ 0 & 1 & t_2 \\ 0 & 0 & 1 \end{pmatrix} \begin{pmatrix} 1 & \bar{1} & 0 \\ 1 & 0 & 0 \\ 0 & 0 & 1 \end{pmatrix} \begin{pmatrix} 1 & 0 & \bar{t_2} \\ 0 & 1 & t_2 \\ 0 & 0 & 1 \end{pmatrix} = \begin{pmatrix} 1 & \bar{1} & 1 \\ 1 & 0 & 0 \\ 0 & 0 & 1 \end{pmatrix}$$

and

$$\begin{pmatrix} 1 & 0 & t'_1 \\ 0 & 1 & t'_2 \\ 0 & 0 & 1 \end{pmatrix} \begin{pmatrix} 0 & \bar{1} & 0 \\ 1 & \bar{1} & 0 \\ 0 & 0 & 1 \end{pmatrix} \begin{pmatrix} 1 & 0 & \bar{t'_1} \\ 0 & 1 & \bar{t'_2} \\ 0 & 0 & 1 \end{pmatrix} = \begin{pmatrix} 0 & \bar{1} & 1 \\ 1 & \bar{1} & 0 \\ 0 & 0 & 1 \end{pmatrix}$$

for rotation, respectively, by 60° and 120° followed by translation by $\vec{a}$. By matrix multiplication

$$\begin{pmatrix} 1 & \bar{1} & t_2 \\ 1 & 0 & t_2-t_1 \\ 0 & 0 & 1 \end{pmatrix} = \begin{pmatrix} 1 & \bar{1} & 1 \\ 1 & 0 & 0 \\ 0 & 0 & 1 \end{pmatrix}$$

and

$$\begin{pmatrix} 0 & \bar{1} & t_1'+t_2' \\ 1 & \bar{1} & 2t_2'-t_1' \\ 0 & 0 & 1 \end{pmatrix} = \begin{pmatrix} 0 & \bar{1} & 1 \\ 1 & \bar{1} & 0 \\ 0 & 0 & 1 \end{pmatrix}$$

from which $t_1=t_2=1$ (the implied 60° rotation is about the lattice point at $x=y=1$) and $t_1'=2/3$, $t_2'=1/3$ (the implied 120° rotation is about the point $x=2/3$, $y=1/3$).

Bibliography

1. L. Landau and E. Lifshitz, Statistical Physics, Pergamon Press, New York, p. 430-456 (1959).

2. H. F. Franzen and J. A. Merrick, J. Solid State Chem. $\underline{33}$, 371 (1980).

3. T. J. A. Popma and C. F. van Bruggen, J. Inorg. Nucl. Chem. $\underline{31}$, 73 (1969).

4. R. D. Heyding and L. D. Calvert, Can. J. Chem. $\underline{35}$, 449 (1957).

5. H. F. Franzen and T. J. Burger, J. Chem. Phys. $\underline{49}$, 2268 (1968).

6. V. Heine and J. D. C. McConnell, Phys. Rev. Lett. $\underline{46}$, 1092 (1981).

7. H. F. Franzen and B. C. Gerstein, A.I.Ch.E.J. $\underline{12}$, 364 (1966).

8. J. Neubuser and H. Wondratschek, "Maximal subgroups of the space-groups", University of Karlsruhe, internal publication.

9. N.F.M. Henry and K. Lousdale, ed., International Tables for X-ray Crystallography, International Union of Crystallography, Kynoch Press, Birmingham, England (1969).

10. H. F. Franzen, C. Haas and F. Jellinek, Phys. Rev. B. $\underline{10}$, 1248 (1974).

11. C. Haas, Phys. Rev. $\underline{140}$, A 863 (1965).

Inorganic Chemistry Concepts

Editors: C. K. Jørgensen, M. F. Lappert,
S. J. Lippard, J. L. Margrave, K. Niedenzu, H. Nöth,
R. W. Parry, H. Yamatera

Volume 1
R. Reisfeld, C. K. Jørgensen

Lasers and Excited States of Rare Earths

1977. 9 figures, 26 tables. VIII, 226 pages
ISBN 3-540-08324-3

Volume 2
R. L. Carlin, A. J. van Duyneveldt

Magnetic Properties of Transition Metal Compounds

1977. 149 figures, 7 tables. XV, 264 pages
ISBN 3-540-08584-X

Volume 3
P. Gütlich, R. Link, A. Trautwein

Mössbauer Spectroscopy and Transition Metal Chemistry

1978. 160 figures, 19 tables, 1 folding plate.
X, 280 pages. ISBN 3-540-08671-4

"...The book is thus a remarkable source of information not only for aspiring research students but for any people concerned with physics and chemistry research in university and industry. It should remain an important reference for a long time."
Die Naturwissenschaften

Volume 4
Y. Saito

Inorganic Molecular Dissymmetry

1979. 107 figures, 28 tables. IX, 167 pages
ISBN 3-540-09176-9

"...The book is directed towards a general and synthetic understanding of chiral molecules, and their unique property of optical activity, in the field of transition metal chemistry. The level of treatment is suited to graduate or advanced undergraduate teaching. For these roles, and for library reference, the book is strongly recommended."
Nature

Volume 5
T. Tominaga, E. Tachikawa

Modern Hot-Atom Chemistry and Its Applications

1981. 57 figures, 34 tables. VIII, 154 pages
ISBN 3-540-10715-0

This book has long been awaited by students and researchers seeking a clear introduction to the concepts of modern hot atom chemistry. Various applications to inorganic, analytical, geochemical, biological, and energy-related studies are discussed with a view toward the promotion of interdisciplinary collaboration. Topics of current interest, such as NEET, laser isotope separation and mesic chemistry, are also described to expand the scope for future development in hot atom chemistry.

Volume 6
D. L. Kepert

Inorganic Stereochemistry

1982. 206 figures, 45 tables. XII, 227 pages
ISBN 3-540-10716-9

An important recent advance concerns the stereochemistry of molecules containing ring systems, which are extremely important throughout chemistry. Such molecules may not have stereochemistries corresponding to any of the usual polyhedra, but are intermediate between two different idealized polyhedra. The precise location of a particular molecule along this continuous range of stereochemistries depends upon the geometric design of the ring system, which includes the number of atoms in ring and the size of these atoms.
The simple techniques outlined in this work are the best way, and in most cases the only way, that such complicated structures with coordination numbers from four to twelve can be predicted.

Volume 7
H. Rickert

Electrochemistry of Solids

An Introduction
1982. 95 figures, 23 tables. XII, 240 pages
ISBN 3-540-11116-6

The electrochemistry of solids is of great current interest to research and development. The technical applications include batteries with solid electrolytes, high-temperature fuel cells, sensors for measuring partial pressures or activities, display units and, more recently, the growing field of chemotronic components. The science and technology of solid-state electrolytes is sometimes called solid-state ionics, analogous to the field of solid-state electronics. Only basic knowledge of physical chemistry and thermodynamics is required to read this book with utility. The chapters can be read independently from one another.

Springer-Verlag
Berlin Heidelberg New York

V. N. Kondratiev, E. E. Nikitin

Gas-Phase Reactions

Kinetics and Mechanisms

1981. 1 portrait, 64 figures, 15 tables. XIV, 241 pages
ISBN 3-540-09956-5

Contents: General Kinetic Rules for Chemical Reactions. – Mechanisms of Chemical Reactions. – Theory of Elementary Processes. – Energy Exchange in Molecular Collisions. – Unimolecular Reactions. – Combination Reactions. – Bimolecular Exchange Reactions. – Photochemical Reactions. – Chemical Reactions in Electric Discharge. – Radiation Chemical Reactions. – Chain Reactions. – Combustion Processes. – References. – Subject Index.

The science of contemporary gas kinetics owes much to the pioneering efforts of V. N. Kondratiev. In this book, he and his coauthor E. E. Nikitin describe the kinetics and mechanisms of gas reactions in terms of current knowledge of elementary processes of energy transfer, uni-, bi- and trimolecular reactions.
Their consideration of formal chemical kinetics is followed by a discussion of the mechanisms of elastic collisions, and of unimolecular, combination and bimolecular reactions. In addition, they have devoted several chapters to the kinetics of the more complicated photochemical reactions, reactions in discharge and radiation-chemical reactions, the general theory of chain reactions, and processes in flames. Particular attention is paid to non-equilibrium reactions, which occur as a result of the Maxwell-Boltzmann distribution principle.
This comprehensive and critical presentation of gas phase kinetics will prove an excellent source of information for chemists and physicists in research and industry as well as for advanced students in chemistry and chemical physics (540 references).

Springer-Verlag
Berlin
Heidelberg
New York

Lecture Notes in Chemistry